DECIDING TO INNOVATE

DECIDING
TO INNOVATE
How Firms Justify
Advanced Technology

JAMES W. DEAN, JR.

BALLINGER PUBLISHING COMPANY
Cambridge, Massachusetts
A Subsidiary of Harper & Row, Publishers, Inc.

International Standard Book Number: 0-88730-189-4

Library of Congress Catalog Card Number: 87-17432

Printed in the United States of America

Library of Congress Cataloging-in-Publication Data

Dean, James W., 1956.
  Deciding to innovate.
  Includes bibliographies and index.
    1. United States — Manufactures — Technological
innovations — Decision making — Case studies.
2. Technological innovations — Management — Case studies.
I. Title.
HD9725.D35   1987   658.5'14   87-17432

ISBN 0-88730-189-4

To my parents,
Warren and Helen Dean,
with love

CONTENTS

LIST OF FIGURES

LIST OF TABLES

PREFACE

This book is about how modern American corporations make decisions about advanced manufacturing technology (AMT). It is based on research I conducted in five manufacturing organizations over a three-year period. This research grew out of a general concern that American manufacturing organizations were gradually losing their ability to compete in world markets because of a failure to implement state-of-the-art technologies. This failure is often due to an inability to justify investment in technology, using traditional techniques.

As I look back on my experiences with these five firms, I am more convinced than ever that applying the traditional justification process to AMT often results in a serious impediment to our competitiveness. Two qualifications, however, should help put this book in proper perspective. First, the decision to adopt a particular technology is only the first step in becoming or remaining competitive. In order to reap the benefits of technology, it must be successfully implemented. Poor implementation is often worse than not adopting a technology in the first place. Second, technology per se is not a panacea, even when well implemented. It is only when technology is combined with effective management that firms will be able to compete in the increasingly rigorous international marketplace. In other words, while justification of advanced technology is a necessary condition for competitive success, it is far from sufficient.

The book is divided into three parts. Part I provides the context for my research. Chapter 1 describes the competitive conditions that are faced by American manufacturing corporations, the potential of AMT to help them succeed in this environment, and an introduction to the justification process. Chapter 2 contains a description of some of the fundamental findings of other research on the nature of the innovation decision process; thus my findings on the AMT justification process can be placed in the context of previous research. Chapter 3 describes my research methods and provides an introduction to the cases.

Part II includes the cases I wrote as a result of my study in each of five firms. In order to conceal their identities, the names of the firms are changed, and certain details are disguised. Each case centers on the justification process for a particular type of AMT. Chapters 4 through 8 present the individual cases for International Metals, Defense Technology, Temple Laboratories, American Plumbing Fixtures, and Monumental Building Supply, respectively.

Part III brings together the findings from the cases into an overall model and derives implications from these findings for both management and research. The findings are presented in detail in Chapter 9; their implications are discussed in Chapter 10. Chapter 10 also includes a comparison of my findings to those of previous works on innovation decisions, as well as a list of recommendations on how managers can use the justification process to increase the chances that AMT will be approved.

The book is intended for both researchers and managers. I have attempted throughout to make the material relevant and understandable to both groups. Since these groups have different interests, however, some parts of the book will probably appeal more to one than the other. Researchers interested primarily in the innovation decision process might be less interested in Chapter 1's description of the business context of this process, while managers may want to skip the discussion of the research literature on innovation decisions and parts of the methodological description found in Chapters 2 and 3, respectively. All of the cases, however, will be equally relevant for both groups and can be read independently, in any order. The presentation of findings in Chapter 9 is also important to both groups. Chapter 10 is divided into sections providing implications for both managers and researchers.

It would have been impossible to complete this book, and the project described in it, without the help and support of many people. I would like to thank Dean Eugene Kelley and Associate Dean Paul Rigby for their financial support via the Penn State College of Business Administration Research Projects Fund. I also appreciate the assistance of Val Martin, who has patiently and competently administered my grants for several years.

This project has required a great deal of typing, including several hundred single-spaced pages of interview transcripts, numerous case revisions, and innumerable letters. Many thanks to all of the people who have helped with one or more of these tasks, especially Cynthia Batson, Andrea Delabruere, Mary Greeley-Beahm, Laura Frye, Barbara Lippincott, Teri Macaluso, Shirley Rider, Judy Sartore, Melinda Smith, Janet Smith, and Mary Towner.

I would also like to thank Gerry Susman, who inadvertantly sparked my interest in this project while we were having lunch at a now-defunct Greek pizza place. Gerry contributed to this effort both by getting me interested in the organizational issues surrounding AMT and by organizing the Penn State Automation Forum, which greatly facilitated my understanding of these issues.

Marjorie Richman has been a consistently supportive, helpful, and enthusiastic editor. I especially appreciate her willingness to believe in my project when it was in its early stages. Her suggestions as to the form of the manuscript were also quite helpful.

I owe a particular debt of gratitude to the people from the firms where I conducted this research. Without their unselfish willingness to spend many hours explaining their justification processes to me, I never would have been able to complete this work. I regret that I can only use their pseudonyms here. But to Paul Lutchko, Tom Kidwell, Pete Jordan, George Coyle, and Ernie Howard, and all their colleagues, my sincere thanks.

My most consistent support throughout this work has been from the members of my family, who have shown me a great deal of patience and love. They held down the fort for many days as I disappeared over the mountains to interview someone or other; they then spent many months watching me stare at a little computer screen. They were wise enough to know when I needed to work, and when I needed to be taken away from it. To my wife Jan, and my daughters Noelle and Bridget, I am happy to report that I'll be able to take a few more walks now, and have more time to go outside and play. It is to you I offer my deepest thanks.

James W. Dean, Jr.
University Park, Pennsylvania
March 1987

INTRODUCTION

1 THE COMPETITIVE ENVIRONMENT FOR MANUFACTURING

Even the most casual observer of American business would have noticed that life has gotten tougher for U.S. manufacturing firms over the past decade. Foreign competition has increasingly endangered the profitability — and even the survival — of many American firms. What we refer to as foreign competition is, of course, part of the larger trend of global competition, or "globalization," that has transformed industry in the second half of the twentieth century. In order for a manufacturing firm to be competitive today, it must be able to withstand competition from around the globe. Displaced workers from industries such as textiles in the South, steel in Pennsylvania, automotive in the Midwest, and even the high-tech companies of Silicon Valley can attest to the fact that many of our industries have not withstood the competitive test.

Globalization has engendered several trends that have dramatically changed and intensified the nature of competition for American manufacturing firms. These trends include increased cost competition and emphasis on quality, as well as market fragmentation and shortened product life cycles. Furthermore, they will increasingly define the new ground rules for competition and survival in the markets of the 1980s and 1990s, and will continue to do so into the next century.

CHANGING TRENDS IN MANUFACTURING

Cost Competition

The increasing intensity of cost competition among many American firms is partially a result of the demonstrated ability of foreign producers to deliver products to the United States at prices lower than domestic producers can offer. The current leveling off of inflation, or even the deflation of the American economy that precludes the passing on of high production costs to consumers (Coopers and Lybrand 1986), has contributed to cost competition. In an environment where prices are flat or even declining and firms are scrambling for market share, cost-cutting is one of the few means available to maintain profit margins.

Organizations have responded to this increased pressure with layoffs of unprecedented size in both blue- and white-collar areas, as well as with cost-cutting measures in virtually every area of operation. Between 1970 and 1984, the Fortune 500 permanently lost between four and six million jobs (Drucker 1985). Firms have also been quick to pass along cost pressures to their suppliers. For example, American color television manufacturers, encountering strong competition from Japan, have demanded substantial price reductions from the glass producers that provide them with screen panels and picture tube funnels.

Quality

Perhaps the most noteworthy recent development in manufacturing is the increased emphasis on quality. Throughout the 1970s and 1980s, American consumers have been abandoning domestic producers in favor of the higher quality products of their foreign competitors, notably the Japanese. This has been particularly true for such high value-added products as automobiles and consumer electronics. As consumers have come to expect the level of quality offered by Japanese and other competitors, American firms have undertaken massive programs to attain the level of product quality that will allow them to survive. In most industries, the question is not *whether* a firm has implemented a quality improvement program, but *which* one or ones is it using. Many companies, including Ford Motor Company, have featured their quality efforts prominently in their advertising.

This newfound emphasis on quality, like the increased cost pressures, has increasingly spread to suppliers. Many firms have recognized that if component quality is low, the final product cannot be of high quality. Suppliers are under tremendous pressure to raise their quality standards. Meetings between a firm and its suppliers to discuss quality have become more common. Firms have started to eliminate inspection of incoming materials from high-quality suppliers in whom they can place confidence. Thus both the importance and the minimum acceptable level of product quality have substantially increased in both consumer and industrial markets.

Flexibility

Over the past decade, flexibility has joined cost and quality in the lexicon of American manufacturing objectives. The greater need for flexibility stems from two trends: shortened product life cycles and market fragmentation. Firms are increasingly unable to produce a standard product over a long period of time; a rapidly evolving mix of customized products is in demand. Some observers, such as Piore and Sabel (1984), feel that manufacturing organizations in the United States and Europe will be forced to make "flexible specialization" a principal strategic objective in the years ahead.

Market fragmentation already characterizes many industries, including apparel, building materials, bread, medical products, office equipment, steel, and tools (Sabel 1982; Hirschhorn 1984). Businesses in North America, Europe, and Japan expect to be making even more customized products over the next few years (Ferdows et al. 1985). The tendency of many firms to produce for a variety of markets contributes to this fragmentation, as does the increasing diversity of lifestyles within the American culture.

A number of observers have noted the increasing brevity of product life cycles (Skinner 1978; Jelinek and Goldhar 1984; Ferdows et al. 1985). One reason for this trend is the potential for technological breakthroughs to create "dematurity" in a market (Abernathy, Clark, and Kantrow 1983). Dematurity occurs when new technology so affects the design or manufacture of a product that the market dynamics approach those of a new product.

This combination of shortened product life cycles and market fragmentation places a substantial premium on flexibility in manufacturing. In order to compete in many industries today, and even more in the future, firms will need the ability to produce a wide variety of customized products simultaneously and to abandon production of current products in favor of new ones quickly. Globalization has truly rewritten the rules of competition: in order to be viable competitors, firms must be able to manufacture a rapidly changing mix of high-quality, customized products at very low costs.

In the face of these trends, our trade imbalance has grown ever larger, and industry lobbyists are flocking to Capitol Hill pleading for protection. Some domestic industries, such as footwear, have essentially abandoned manufacturing, performing only final assembly on products that are made overseas and imported in parts, thus circumventing legislated import limits. Others, consumer electronics companies, for example, have gone even further, merely putting their names on videocassette recorders and other products made in Japan. This concession of our core manufacturing activities to offshore firms, while perhaps generating short-run financial returns, has created what *Business Week* ("The Hollow Corporation" 1986) has termed "the hollow corporation."

WHY HAVE WE FAILED?

Many possible causes for our competitive failures have been proposed. Lower labor costs, more work-oriented cultures, unfair trade practices, and government subsidies have all been suggested as explanations for the ability of other countries, especially Japan, to defeat American firms consistently in international competition.

There is, no doubt, some truth to these contentions, yet one gets the unpleasant feeling that our wounds are at least partially self-inflicted. In fact, a voluminous literature identifying the competitive shortcomings of American firms has accumulated (e.g., Lawrence and Dyer 1983; Abernathy, Clark, and Kantrow 1983). Suggestions for improving our competitive position from within have included, in addition to outright protectionism, taking a longer term perspective as managers and investors (Hayes and Abernathy 1980) and using human resources more effectively (Ouchi 1981).

Perhaps the most consistent theme that has run through the discussion on the decline in American competitive ability, however, is our failure to remain technologically competitive. American firms have historically been slow to implement state-of-the-art technologies, even though many of them have been developed here. This has often cost them dearly, as firms from other countries have utilized these technologies. In so doing, they have created a wide technological gap, which in turn resulted in disparities in cost, quality, and flexibility. The most recent example of this phenomenon is the reluctance of American firms to implement a set of technologies collectively known as advanced manufacturing technology, which may be our last hope for global manufacturing viability in many industries. We now turn to a description of AMT and its potential to deal with the global competitive realities that have been discussed.

ADVANCED MANUFACTURING TECHNOLOGY

Advanced manufacturing technology, or AMT, is a set of microprocessor-based technological innovations that have the potential to revolutionize manufacturing (Bylinsky 1981). Some of its major applications include computer-aided manufacturing, computer-aided design, computer-aided engineering, manufacturing resources planning, group technology, and computer-integrated manufacturing. Each of these is briefly described below.

Computer-Aided Manufacturing

Computer-aided manufacturing (CAM) consists of robotics, numerically controlled machines, computer numerically controlled machines, direct numerically controlled machines, and flexible manufacturing systems. CAM is the most hardware-intensive of the advanced technologies and has the most direct contact with the fabrication and assembly of products. Robots are defined by the Robotic Industries Association as "reprogrammable, multifunction manipulators designed to move materials, parts, tools, or specialized devices through variable programmed motions for the performance of a variety of tasks without human intervention" (Fisk 1981). The essential components of a robot are a power source, a computerized controller, and a manipulator. Robots have been used in a number of industries for such applications as painting, welding, and material handling.

Numerically controlled (NC) machines "read" coded paper tapes that contain instructions for performing operations such as cutting or grinding. Computer numerical control (CNC) and direct numerical control (DNC) machines are extensions of NC whereby computers, either alone or in hierarchical networks, replace the coded paper tapes. CNCs and DNCs make it easier to reprogram machine tools by either editing existing programs or creating new ones. When CNC machines are combined with material handling equipment, the result is a flexible manufacturing system (FMS). This type of system consists of "groups of machine tools served by automated tool-and-workpiece transport and handling equipment, all operating under hierarchical computer control to produce broad families of machined parts" (Merchant 1983: 36).

The most important feature of CAM is its flexibility, made possible by its ability to be reprogrammed to perform a variety of operations. Another important feature is its ability to repeat endlessly the same operations in an essentially identical manner. CAM can also maintain machining tolerances (sometimes in the .001-inch range) that human beings cannot. As a result, product performance and reliability are increased. CAM generally performs at lower costs than the direct labor it replaces.

Computer-Aided Design

Computer-aided design (CAD) replaces the draftsman's or engineer's table and T-square with a computer terminal. CAD "increases the productivity of design engineers by doing routine tasks for them, such as sorting and searching through plans, labeling, and, of course, drafting" (Rosenbaum 1983: 49). It can develop instructions for use by CNC machines. Design software calculates the most efficient tool paths that a machine can use in producing whatever has been designed. The instructions that result from such calculations are then uploaded to the machine tool controller, and the newly designed part is produced without further human intervention. This capability is characteristic of many of the systems developed in the last few years.

Computer-Aided Engineering

Computer-aided engineering (CAE) allows engineers to test their designs without building expensive prototypes. For example, CAE systems can demonstrate the effects of stress or heat on a given product design. Such systems are used in the electronics industry to model the functionality of printed circuit board designs. Both CAD and CAE can reduce product development time considerably. They also make it easier to design customized products, thus reducing cost and increasing flexibility.

Manufacturing Resources Planning

Manufacturing resources planning (MRP) encompasses as well as supersedes materials requirements planning; for this reason it is often referred to as MRP II. MRP performs a variety of functions that are related to the scheduling and regulation of production in the factory and is typically comprised of modules. Standard MRP modules are master production schedule, bill of materials, shop floor control, and materials requirements planning. From data about what products need to be completed within a certain time period, and the raw materials, manufacturing processes, and labor needed to produce them, MRP generates among other things, a production schedule and information about necessary raw materials orders. Many firms have used MRP to reduce inventory and throughput time and to increase the reliability of their deliveries.

Group Technology

Group technology (GT) operates on a system whereby parts are grouped together into families, thus taking advantage of their similarities in design and/or manufacturing (Groover 1983). Like MRP, GT is actually more of a philosophy or practice than a technology per se. Given the complexity of most manufacturing operations, however, neither GT nor MRP could be implemented without the availability of sophisticated software packages and powerful computer hardware.

Group technology provides advantages in both design and manufacturing. On the design side, engineers can avoid reinventing parts by examining the GT database to see if an existing design could be modified. In manufacturing, GT can be used to organize the factory floor into "manufacturing cells," each of which concentrates on a particular type of part family. This arrangement eliminates much of the moving and queue time that characterizes the traditional factory and thus dramatically reduces throughput time.

Computer-Integrated Manufacturing

Computer-integrated manufacturing (CIM) is difficult to define. It has been variously described as "a philosophy," "a way of life," and "a journey, not a destination." Fundamentally, CIM represents the integration of many or all of the elements of AMT. It is not a particular machine or piece of software, nor even a particular combination of such elements. CIM is born of the recognition that while any automated device may provide advantages for manufacturing and/or design, truly substantial benefits will be realized only when these devices and systems are integrated into a coherent system. It thus denotes in one sense the computer architecture that coordinates and controls an automated factory, and in another sense the integrated system as a whole. CIM cannot currently be bought as a "turn-key" system (although many of the elements may be purchased), and it is unlikely that exactly the same CIM configuration could be utilized by different organizations.

AMT IN AMERICAN INDUSTRY

As we can see, the various forms of AMT have tremendous potential for enhancing the competitive capacity of American manufacturing firms. They offer strategic advantages in cost, quality, and flexibility, precisely those areas on which global competition has placed a premium. AMT has provided tremendous benefits to firms such as General Electric, IBM (Saporito 1986), and Allen-Bradley (Bylinsky 1986). Given this, one would expect manufacturers to be rushing headlong to adopt some or all of these technologies. Outside of a few industries such as automobiles, aerospace, and electronics, however, this is not the case.

International Data Corporation, a market research organization, estimates that only about a dozen companies have attempted to utilize CIM in their facilities, and fewer than 250 American firms have done more than a few small-scale projects using AMT, according to a *Business Week* special report on the issue (Brandt and Port 1986). This report also notes that most AMT spending comes from about 2 percent of manufacturing firms in this country and that the United States currently holds only an increasingly narrowing edge over international competitors in CIM.

THE JUSTIFICATION PROCESS AND AMT

A primary hindrance to the adoption of AMT is the justification process by which firms calculate the financial return to be expected from AMT (or any other capital investment) and decide whether to devote funds to it. Measures commonly used to make such decisions include payback period (the amount of time necessary to break even on the investment), return on investment, internal rate of return, and net present value. Firms generally have minimum cutoff values, or "hurdle rates," for the measures they use, for example, a two-year payback or 20 percent internal rate of return. Proposed investments that do not exceed the hurdle rates are usually not funded.

In many cases, the justification process determines that AMT does not provide adequate return to justify its cost, and thus it is not pursued. As we shall see, this outcome is often a function of the intangibility and long-term nature of AMT's benefits, as compared with the very tangible and substantial short-term expenditures required for initial investment in AMT. In fact, a negative outcome is sufficiently common that Huber (1985: 46) has claimed, "Traditional justification procedures... are quite likely the single greatest barrier to the utilization of new manufacturing technologies by U.S. manufacturing industries." Thus we are left with a situation in which technologies that hold enormous potential for revitalizing American manufacturing are being neglected due to their failure to meet financial justification criteria.

A number of criticisms have been leveled at traditional justification techniques, both for capital investments in general and AMT in particular (Blumberg and Gerwin 1984). First, these techniques are inherently biased toward investments that will pay back in the short run and against those that offer even substantial long-run returns. Second, they are based on a financial, rather than strategic, conception of business needs. Overreliance on traditional justification techniques represents "management by the numbers," a philosophy whereby managers do not concern themselves with the underlying dynamics of the industries in which they compete, but instead focus exclusively on monitoring and controlling certain key financial indicators. Third, traditional justification for technology focuses to an inappropriate degree on reduction in labor cost, even as labor represents an ever-smaller percentage of overall product cost.

Many AMT proponents believe that the traditional justification techniques are particularly inappropriate for evaluating AMT, as they materially underestimate the benefits to be derived from investment in advanced technology. Due to its flexibility, AMT will often have a useful life well in excess of the product line for which it is initially adopted. Yet traditional justification practices consider only benefits to be derived from a current product, or perhaps one in the immediate future. AMT can also be distributed among several product lines as demand shifts, a value which is generally seen as impossible to quantify and therefore not included in ROI calculations (Kaplan 1986). In addition, in order to control costs and keep decisions at lower levels, AMT champions in organizations often present individual elements of AMT for separate approval over a period of time. Unfortunately, justifying elements of AMT individually ignores the synergy that is created when several elements are integrated into a system.

It is also difficult to measure the benefit of organizational learning that results from the implementation of new technology, a benefit that increases the ROI of future AMT projects. This and other similar factors are often referred to as "intangibles," that is, intangible benefits of investment in advanced technology. According to Brandt and Port, "Nearly all financial yardsticks ignore the benefits of such intangibles as flexibility, better quality, and shorter lead times in getting new products to market. Yet these intangibles provide the most concrete reasons for automating" (1986: 74). Intangible benefits stemming from investment in CIM are identified in a case study conducted by Susman and Dean (1986). Steven Klabunde, vice president for manufacturing systems at Giddings & Lewis, Inc., commented on the dilemma of trying to evaluate such intangible factors as quality improvements that result from AMT. His comment summarizes the feelings of many managers who have experienced frustration in AMT justification:

> It is impossible to put precise figures on the benefit of better quality, yet you know it can be invaluable to you in certain markets. The benefit lies, for example, in less scrap and salvage since you're making the product right the first time. You'll have lower warranty costs, fewer service calls and lower service costs. As your product's reputation for quality becomes known in the marketplace, your share of the market will increase...the reputation for quality can be accompanied by higher prices and higher margins. There is no way you can set down a meaningful dollar figure for those values, but they are nonetheless real. It may be necessary to use some intuition and just assess the degree to which they are of value and will enhance your company's future. But, surely, your inability to put a figure beside each [benefit] on some justification form doesn't mean they can be ignored. To do so would be to preclude major steps forward in manufacturing technologies and efficiencies. And the jeopardy is that a competitor of yours will be doing what you decided not to do (Huber 1985: 48–49).

Perhaps the fundamental issue is one of survival. AMT proponents argue that investment in technology is an outright necessity for survival in many industries; not to invest in it for any reason is simply a mistake. J. Tracey O'Rourke, president of Allen-Bradley — a subsidiary of Rockwell International that has gained notoriety for its CIM implementation — has said that, in many industries, there will be survivors and nonsurvivors, and that if a firm is not among the survivors, "ROI doesn't mean a damn" (Bylinsky 1986: 65).

Of course, such comments criticizing capital budgeting techniques are part of a larger discussion concerning the place of manufacturing in a business enterprise. A number of scholars (e.g., Skinner 1978) have argued that manufacturing's subordination to finance and marketing is at the core of our competitive difficulties and that the inability to justify capital spending to upgrade manufacturing is just a symptom of the larger problem. Because of its importance and potential, AMT has become a lightning rod for discussion of the role of manufacturing in competitive strategy.

Given the shortcomings of traditional justification processes for AMT investment, how have firms reacted? The senior executives and financial officers of most firms clearly do not share the sentiments expressed above. Most companies have simply retained their traditional approaches, demanding that AMT proponents find a way to express its benefits in these terms. Others have augmented their traditional techniques to provide an opportunity to at least list the intangible or qualitative benefits that can be expected from AMT (Kaplan 1986). A small number of firms have adopted or developed new techniques for justification, or abandoned traditional techniques entirely in favor of justification based on strategy or survival.

Regardless of which of these paths they have taken, however, managers are often very uncertain about whether they really need AMT. While aging facilities are increasingly unable to produce goods of acceptable quality and cost, the necessary investment in automation is enormous, especially where CIM, the current state-of-the-art technology, is concerned. And success is anything but guaranteed; failure stories abound.

In this atmosphere of uncertainty, justification decisions shift away from emphasizing the use of numbers toward the organizational and interpersonal dynamics involved in the process. As Goldhar (1986) has pointed out, justification is not an accounting problem, it's a behavioral problem. As the facts surrounding an issue or decision become more ambiguous and less deterministic, the vacuum they leave in terms of influencing a decision is rapidly filled by social dynamics, including politics. In this environment, whether an organization adopts such crucial technologies as CAD or CIM may have less to do with their IRRs (since no one is really certain what they are) than with the interpersonal dramas that constitute the justification process. Since this process takes place in and is affected by a given organizational context, such factors as organizational structure and culture will play an important role in firms' decisions about advanced technology.

The purpose of this book is to examine in depth how the justification process unfolds. Using five firms as examples, I aim to identify the financial, organizational, and interpersonal barriers that exist in the decisionmaking process that constitutes justification. This investigation will attempt to provide insight into how firms have ignored, hurdled, or crashed through these barriers to move toward AMT. By presenting the justification strategies used by successful firms, I hope to offer options to firms that are in the process of AMT justification.

Obviously, the conditions that have given rise to the justification dilemmas described in this book are temporary. Ten years ago, AMT existed mainly on the drawing board; ten years from now, experience with AMT across many industries and organizations will render its justification routine. I hope that the work reported here will help to document how some firms were able to justify the investment in AMT that will allow it to eventually become routine.

A number of broader issues, however, surround the adoption of a particular type of technology by a particular set of firms at a particular time. Technological innovation will continue to occur, even when technologies that are currently dazzling are considered antiquated. Thus this study should shed some light on how decisions about *any* technological innovation are made by business organizations, be it computerized manufacturing or the use of artificial intelligence to design products or make strategic decisions.

Even more broadly, decisions characterized by uncertainty are the fabric of managerial life. Clear-cut decisions are to an increasing degree nonevents for managers; these are designed into a company's software and executed by low-level employees. Important yet uncertain decisions, on the other hand, often involve many actors with different backgrounds and objectives, all of whom are trying in various ways to help their firm survive and prosper in a changing world. AMT justification decisions are, in many ways, a prototype of such organizational decisions—they are urgent, ambiguous, and unforgiving.

REFERENCES

Abernathy, W.J., K. B. Clark, and A.M. Kantrow. 1983. *Industrial Renaissance: Producing a Competitive Future for America.* New York: Basic Books.

Blumberg, M., and D. Gerwin. 1984. "Coping with Advanced Manufacturing Technology." *Journal of Occupational Behaviour* 5, no. 2 (April): 113-30.

Brandt, R., and O. Port. 1986. "How Automation Could Save the Day." *Business Week,* March 3, pp. 72-74.

Bylinsky, G. 1981. "A New Industrial Revolution Is on the Way." *Fortune* 104, no. 7 (October 5): 106-14.

———. 1986. "A Breakthrough in Automating the Assembly Line." *Fortune* 113, no. 11 (May 26): 64-66.

Coopers and Lybrand. 1986. *Annual Report on the Worldwide Economic and Business Climate.* Washington, D.C.

Drucker, P.F. 1985. *Innovation and Entrepreneurship: Practices and Principles.* New York: Harper and Row.

Ferdows, K., J.G. Miller, J. Nakane, and T.E. Vollman. 1985. "Evolving Manufacturing Strategies in Europe, Japan, and North America." Unpublished paper, Boston University.

Fisk, J.D. 1981. *Industrial Robots in the United States: Issues and Perspectives.* Report No. 81-78E. Prepared for the U.S. Congressional Research Service. Washington, D.C.: U.S. Government Printing Office.

Goldhar, J.D. 1986. "Strategic Implications of Advanced Manufacturing Technology." Paper presented at the Conference on the Justification and Acquisition of New Technology, University of Rochester Graduate School of Management, Rochester, N.Y., September 18-19.

Groover, M. 1983. "Fundamental Operations in Data-Driven Automation." *IEEE Spectrum,* Special Issue, 20, no. 5 (May): 65-69.

Hayes, R.H., and W.J. Abernathy. 1980. "Managing Our Way to Economic Decline." *Harvard Business Review* 58, no. 4: 67-77.

Hirschhorn, L. 1984. *Beyond Mechanization: Work and Technology in a Postindustrial Age.* Cambridge, Mass.: MIT Press.

"The Hollow Corporation." 1986. *Business Week,* Special Report, March 3, pp. 56-86.

Huber, R.F. 1985. "Justification: Barrier to Competitive Manufacturing." *Production* 95, no. 9 (September): 46-51.

Jelinek, M., and J.D. Goldhar. 1984. "The Strategic Implications of the Factory of the Future." *Sloan Management Review* 25, no. 4 (Summer): 29-37.

Kaplan, R.S. 1986. "Must CIM be Justified by Faith Alone?" *Harvard Business Review* 64, no. 2 (March/April): 87-95.

Lawrence, P.R., and D. Dyer. 1983. *Renewing American Industry: Organizing for Efficiency and Innovation.* New York: The Free Press.

Merchant, M.E. 1983. "Production: A Dynamic Challenge." *IIEE Spectrum,* Special Issue, 20, no. 5 (May): 36-39.

Ouchi, W.G. 1981. *Theory Z: How American Business Can Meet the Japanese Challenge.* Reading, Mass.: Addison-Wesley.

Piore, M.J., and C.F. Sabel. 1984. *The Second Industrial Divide: Possibilities for Prosperity.* New York: Basic Books.

Rosenbaum, J.D. 1983. "A Propitious Marriage: CAD and Manufacturing." *IEEE Spectrum,* Special Issue, 20, no. 5 (May): 49-52.

Sabel, C.F. 1982. *Work and Politics: The Division of Labor in Industry.* Cambridge: Cambridge University Press.

Saporito, B. 1986. "IBM's No-Hands Assembly Line." *Fortune* 114, no. 6 (September 15): 105-109.

Skinner, W. 1978. *Manufacturing in the Corporate Strategy.* New York: John Wiley.

Susman, G.I., and J.W. Dean, Jr. 1986. "A Case Study of Four Projects Supported by the Industrial Modernization Incentives Program of the Department of Defense." Unpublished paper, Pennsylvania State University.

2 THE FOUNDATIONS OF INNOVATION RESEARCH

There are few in-depth examinations of the justification process for AMT on which this study could be built. The technologies and issues that surround AMT are sufficiently new that very little research specifically devoted to the subject and its justification process has been performed to date. This does not mean, however, that one must start from scratch in an examination of the decisionmaking process that surrounds AMT justification. As I indicated in the last chapter, the technologies that comprise AMT are innovations — specifically, technological innovations. AMT can also be thought of as simply one of the many types of projects that pass through the capital budgeting process. Thus previous work on the decisionmaking process for innovations — especially technological innovations — as well as on the capital budgeting process, should provide some clues as to what to expect from AMT justification decisions.

For this chapter, I have tried to identify key ideas found in important works related to advanced technology justification. These ideas provide a foundation upon which AMT research can rest. I intend to highlight these ideas, rather than to provide an exhaustive catalogue of where and when each of them has been used.

SCHÖN: CHAMPIONS

Donald Schön, in a 1963 article in *Harvard Business Review,* popularized an idea that has come to be accepted widely among both managers and students of the innovation process: that innovations need "champions". Schön was interested in the process by which radical new inventions became successful, noting that most significant innovations emerge from large organizations rather than from solitary inventors in their garages.

Based on a study of how radical innovations are produced by organizations, Schön arrived at four key findings. First, he noted that new ideas encounter sharp resistance at the outset. Often these ideas look expensive and unfeasible. Second, they receive vigorous promotion to overcome the initial resistance. Promotion is often undertaken by someone other than the original inventor. Third, the informal rather than the formal organization is used in the promotion of new ideas. Until work has progressed sufficiently to subject the new idea to close organizational scrutiny, researchers utilize "bootlegged" funds diverted from other programs.

Schön's fourth and most well-known finding is that generally one person emerges as the champion of the new idea: "Where radical innovation is concerned, the emergence of a champion is required. Given...underground resistance to change...the new idea either finds a champion or *dies*" (p. 84). Beyond identifying the need for a champion, he also provides a description of what the role requires:

> The champion must be...willing to put himself on the line for an idea of doubtful success. He is willing to fail. But he is capable of using any and every means of informal sales and pressure in order to succeed.... Champions...identify with the idea as their own, and with its promotion as a cause, to a degree that goes far beyond the requirements of their job. Many display persistence and courage of heroic quality (pp. 84–85).

Finally, Schön notes that, in order to play their role successfully, champions must have substantial power and prestige in their organization. They also need extensive knowledge of the informal organization and the ability to work across organizational boundaries.

MAIDIQUE: ENTREPRENEURS AND CHAMPIONS

Stimulated by the work of Schön and others, Maidique (1980) both summarizes and contributes to the line of research on roles involved in the innovation decision process in organizations. While Schön had hinted that there may be some "division of labor" between the invention and promotion of ideas, subsequent work attempted to distinguish among the variety of roles often present in innovation decisions. Maidique identifies the SAPPHO project (Rothwell et al. 1974) as one of the first to make such distinctions.

The SAPPHO project examined successful and unsuccessful innovations in several industries and identified four roles associated with successful innovation. The *technical innovator* makes the major technical contribution to the design or development of an innovation. The *business innovator* is a manager who is responsible for the overall progress of the project. The *product champion* contributes to the innovation by promoting it enthusiastically through critical stages. Finally, the *chief executive,* while not necessarily the chief executive officer of the organization, is at the top of the executive structure of the innovating organization. The business innovator is found by this study to be the most critical component of innovation success. This individual's power, respectability, status, and experience are of crucial importance to the process. The role of product champion also was found to be important, but not as important as the business innovator.

Maidique describes how the various roles associated with product innovation evolve as a function of the stages of business development. In the small firm, the entrepreneur helps to define new products while still maintaining control of the overall organization. In the integrated stage — where a firm has a single product line and a functional organization — technological progress and organizational complexity force the entrepreneur to give up technical definition and sponsorship of most new products. Sponsorship of projects that emerge from the technical community is passed on to others in the management ranks, with the entrepreneur having final approval. Maidique (1980: 68) cites a study by Schwartz (1973), where it was found that "middle managers serve as integrators between technical specialists and top management" in firms at this stage.

The diversified firm is the third stage of development identified by Maidique. In such organizations, dominant businesses (representing more than 70 percent of sales) operate within a functional structure, much like integrated firms. Other businesses are run through product divisions. A new role is needed in such divisions to bridge the gap between the entrepreneur and the champion, thus the "executive champion" emerges. Maidique describes such individuals as "several times removed from the detailed technical definition but without the entrepreneurial clout to be the ultimate sponsor" (p. 70).

BOWER: DEFINITION, IMPETUS, CONTEXT

Joseph Bower's well-known book, *Managing the Resource Allocation Process* (1970), describes his study of how diversified firms allocate capital among business units and projects. He spent several years examining the capital allocation process—focusing specifically on the acquisition of new facilities—in one large firm, which he calls "National Products."

Bower's focus is quite a bit different from Schön and Maidique. First, he is interested in resource allocation rather than in innovation per se. Second, his analysis is confined to large, multidivision business organizations. Third, he is more interested in the description of processes than the identification of roles. Since AMT decisions are, as indicated in Chapter 1, made via the capital allocation process, Bower's findings about how this process operates may well be relevant for understanding AMT justification.

As Bower describes it, resource allocation is largely a "bottom-up" decision process in which proposals from lower levels in the organization slowly wend their way upward through multiple levels of hierarchy, any of which could potentially kill them. If a proposal survives this process, the final decision is made by the board of directors; as Bower points out, however, projects are rarely turned down at this level.

The resource allocation process begins when someone in the organization discovers a discrepancy between actual and expected levels of cost, quality, or capacity. Alternatively, an individual may notice that while cost, quality, and capacity objectives are currently being met, a new investment would improve them significantly. The ensuing process is one of *definition,* the first stage of the decision process. A facility is defined as "a new fixed investment, a time of availability, a capacity, a product scope of specified breadth and quality, and a specified level of production cost" (Bower 1970: 53). The definition of a facility is influenced by the initiating individual's job history and likely future, the product's history, other projects that are being planned concurrently, and how the initiating individual's responsibility for the project is defined.

As the decision process evolves, the project moves from the definition to the *impetus* stage, described as an interpersonal and political process. The impetus to move the project forward stems from the willingness of a manager at the division level to commit himself to an idea produced at lower levels and to attempt to secure approval from above. This willingness is affected by the technical and business aspects of the proposed project, as well as by the manager's perception of the benefits of being right about the project's viability versus the costs of being wrong. The impetus process begins when a manager is willing to stake his reputation on an idea. The impetus process is characterized by project approval at progressively higher levels of the organization. Bower notes that decisions made at the intermediate levels may have the most significant impact on the total decision. The impetus process ends when the project is finally approved (or turned down).

Bower makes two key observations about the impetus stage. First, the description of the project, if not its actual definition, continues to evolve throughout this stage: "Although the project definition may not have changed at all from the concept of its originators 'down in the organization', the requisition justifying the project has been screened, revised, and politically disinfected so that it now tells an attractive story in professional tones" (p. 11). Second, the reputation of the middle managers supporting the proposed investment is important. This issue is best illustrated by a quote from one of Bower's managers. "The key question is 'How much confidence has the management built up over the years in my judgment?' A guy in my position must think this way. He loses his usefulness when he loses the confidence of higher executives in the company. Otherwise his ideas will not be accepted when he goes up" (p. 59).

While the definition and impetus processes involve lower level and middle management personnel, respectively, top management also has an important part to play in these processes. Obviously, top managers have the final say as to whether capital will be invested in any large-scale project. But Bower reports that they seldom exercise this veto power, attributing this to their high level of confidence in their subordinates as well as the near-impossibility of being able to stay abreast of the diverse product technologies and businesses in a diversified firm.

Top management's most important contribution to the resource allocation process is creating the structural context in which such decisions are made. In a sense, therefore, top management's influence on the process takes place before even the lowest level managers are involved. Projects can be conceived, defined, and promoted only within a context that is largely the creation of top management. This context includes the formal organization of the firm as well as the systems of information and control that are used to measure the performance of both business units and individual managers. Bower takes care to point out, however, that some elements of the context — factors of a personal or historical nature, called the "situational context" — are not under the control of top management. While the situational context has an important effect on the resource allocation process, it is almost by definition difficult to generalize.

CARTER: DEPARTMENTAL LOYALTIES AND BIASED PROPOSALS

Eugene Carter's (1971) study of strategic decisions in a small- to medium-sized firm also may have implications for the study of AMT adoption decisions. Carter's study was undertaken to test Cyert and March's (1963) theory of decisionmaking. He studied six decisions in one firm, three of them involved acquisitions, another three dealt with investments. We will focus on the investment decisions, since they appear more relevant for understanding AMT justification. These decisions involved the selection of computers, the construction of a remote computer terminal, and the development of an extremely advanced remote computer terminal.

Much as in Bower's study, these investment decisions were initiated at lower levels in the organization, much lower than where acquisition decisions were initiated. The departments in Carter's "Comcor" company used the president's basic philosophies as a guide in identifying projects they might suggest. While Bower focused on individual stakes in the decisionmaking process, Carter emphasized departmental loyalties. In fact, he concluded that the department's wishes were the first consideration in deciding which projects would be supported; the president's wishes and organizational goals were of secondary concern. (Pettigrew (1973) also discussed departmental loyalties in investment decisions.)

In order to facilitate the acceptance of their proposals, subordinates add some bias to their assessments of the prospects for a project. Carter identifies several factors that determine the extent to which a technical subordinate will bias data:

1. The degree to which success depends on the subordinate's representation;
2. The uncertainty of other information on which the appraisal is being based;
3. The relative ignorance of the subordinate's superior;
4. The superior's perception of the uncertainty and relative importance of technical factors; and
5. The desire of the subordinate to exploit his expertise in the evaluative portion of the decision.

Carter also comments on the number of criteria that are used in evaluating an investment proposal. He concludes that both environmental uncertainty and uncertainty about criteria estimates result in a greater number of criteria being used to evaluate a project. For example, if two alternative proposals appear equal on one criterion, such as cost, additional criteria (e.g., quality, reliability) will be taken into consideration until one alternative clearly dominates the other. Information on an increasing number of criteria is collected until management's need for an acceptable amount of data is satisfied.

Finally, Carter echoes in a somewhat different manner a point made by the other works reviewed above. He says that the project sponsor's record, including qualifications, background, and training, influences the project because it determines the threshold level of benefits needed for project approval; those having poorer records need to demonstrate greater benefits before their projects will be approved. This is one way in which management deals with the inherent uncertainty of estimating a project's benefits.

BURGELMAN: INTERNAL CORPORATE VENTURING

Building on the work of Bower, Maidique, and others, Burgelman (1983) conducted an in-depth study of the management of six new business ventures in a large, diversified organization. This work is pertinent to AMT justification because Burgelman chose to study ventures that utilized new product technology.

Burgelman identified two "core processes" and two "overlaying processes" in the development of internal corporate ventures (ICVs). The core processes are *definition* and *impetus,* as defined by Bower. The overlaying processes are *structural context* and *strategic context*. Structural context, again after Bower, is determined by "organizational and administrative mechanisms put in place by corporate management to implement the current corporate strategy" (p. 229). Strategic context, on the other hand, is the process by which an organization's current strategy is extended to accommodate the new businesses that result from an ICV. The decisionmaking process is described as being primarily bottom-up, with middle management playing a key role.

Burgelman discusses in detail the content of the core processes of definition and impetus. He identifies "linking processes" as one component of definition. Linking involves technical linking — the assembling of information to solve technical problems — and need linking — the matching of new technical solutions to new or poorly served market needs. First-line supervisors in a research area are important to linking. They often perform both technical and need linking; because they are aware of corporate strategy, they are able to put technical developments into corporate perspective.

Product championing is another component of the definition process. While many of the aspects of championing Burgelman reports are familiar, it is important to note the extent to which champions for ICVs need to demonstrate that the "impossible [is] in fact possible" (1983: 232). In other words, because the technology at the core of the venture is novel, a large part of the champion's role is to reassure senior decisionmakers that the product will actually work.

Moving to the impetus process, Burgelman notes, in contrast to Bower, that the ICV process does not rely heavily on formal analytic techniques. He notes that each project is unique; it cannot be judged on the basis of previous project standards. The success of the venture at this stage is thus dependent on the product championing activities. Since the credibility of the champion is, in turn, often essential for successful product championing, managers often take on smaller projects with a high likelihood of success in order to build credibility to be used in selling riskier projects later on.

Organizational championing takes place at a later stage in the venturing process and at higher levels in the organization than does product championing. It involves "the establishment of contact with top management to keep them informed and enthusiastic about a particular area of development" (Burgelman 1983: 238). While product championing involves requests for support for a potential product, organizational championing attempts to convince top management of the need for a developing business and that substantial resources should be committed to make this happen.

Not surprisingly, organizational championing is described as a political activity, whereby a manager commits his judgment and puts his reputation on the line. Burgelman notes that the better organizational champions always made sure that their projects were consistent with current organizational strategy. The "brilliant" ones, however, could sell management on the strategic benefits of new businesses that did not fit well with current strategy. Thus organizational championing "required the rare capacity to evaluate the merit of the proposals and activities of different product champions in strategic rather than technical terms" (p. 238).

KANTER: INTEGRATION AND SEGMENTALISM

The final work I will discuss is Rosabeth Kanter's (1983) book on innovation in American corporations, *The Change Masters*. In that it deals primarily with management or human resource innovations, and only incidentally with technology or the capital budgeting process, Kanter's study is distinct from the others I discussed. In its treatment of the development of innovative ideas in large organizations, however, Kanter's work would appear to be very relevant to understanding the AMT decision process.

One of the central features of Kanter's work is the distinction between the two different types of organizations she labels "integrative" and "segmentalist." While these two terms are initially presented as modes of thinking, she argues that they are also descriptive of the organizational cultures and structures that encourage such thinking.

Integrative organizations minimize conflicts between subunits, develop mechanisms for exchange of information and new ideas across organizational boundaries, take multiple perspectives into account when making decisions, and have a coherent direction for the organization as a whole. They provide mechanisms for finding common ground, even among diverse specialists. In this integrative culture, change is embraced and innovation flourishes.

Segmentalist organizations, on the other hand, are anti-change and inhibit innovation. This mode of organization and thought compartmentalizes issues, problems, and even people, and keeps them isolated from one another. Problems are treated as if they were independent of both their organizational context and other current problems. Departments and levels of hierarchy are walled off from one another, and interaction is not encouraged. As Kanter points out, even innovation can be compartmentalized in segmented organizations by assigning it to the R&D department where no one else has to be concerned with it.

This fundamental difference in organization is at the heart of Kanter's explanation of differences in levels of innovation between organizations. Innovative organizations are characterized by integrative cultures, structures, and thinking, while less innovative organizations display more segmentalist characteristics. This is because, while integrative organizations facilitate the efforts of individuals who attempt to innovate, segmentalist organizations discourage such efforts.

Kanter also provides a process model of innovation based on her research experiences. It is divided into three "waves of activity": problem definition, coalition building, and mobilizing and completion. Problem definition is similar to the definition process as described by Bower and Burgelman. It requires careful listening to available information and translation of vague ideas and assignments into concepts for concrete action. Both technical and "political" data, as well as any other data needed to make a convincing case for the proposed innovation, figure into the problem definition. Kanter's definition of "political" is related to her concepts of integration and segmentalism: "An idea must be sold, resources must be required...and some variable numbers of other people must agree to changes in their own areas — for innovations generally cut across existing areas and have wider organizational ripples, like dropping pebbles into a pond" (p. 216).

Given the notion that innovations often have implications for several organizational subunits, and must be negotiated with those whom they affect, the reason for Kanter's emphasis on the dangers of segmentalism becomes clear. When subunits are kept separate from one another and hostility develops between them, it becomes much more difficult for them to arrive at a consensus about an innovation that will affect all of them. Thus it is more likely that one or more of the affected parties will manage to scuttle the whole process.

Coalition building is the second wave of activity in Kanter's description of the innovation process. This process is necessary to secure both support and resources. It requires getting informal approval from one's boss and "preselling" the idea to one's peers. Kanter notes that higher level managers look for evidence that an idea is supported by its proponents' peers before committing themselves to it. Sometimes "horse trading", that is, offering promises of some type of payoff in the event the project succeeds, is necessary to secure this support. Coalition building also requires "securing blessings" from relevant higher level officials and "formalizing the coalition." The former generally requires formal presentations, as well as standing up under the intense scrutiny to which innovative projects are subjected; each executive is looking at the likelihood of being able to sell the project to his or her own superiors. Formalizing the coalition involves the creation of a task force or advisory committee.

Kanter's third wave of activity, mobilizing and completion, is more a description of the implementation of innovations than of their approval per se. It appears that the line between approval and implementation is more blurred for managerial innovations than for the types of technological innovations that are our subject here. For the sake of completeness, however, mobilizing and completion includes handling opposition and blocking interference, maintaining momentum, secondary redesign, and external communication.

SUMMARY AND COMPARISON

Despite the lack of research on the justification process for AMT, there are clearly a number of well-developed ideas concerning the innovation and resource allocation processes. The conclusions of each of the works reviewed overlap to some extent with the others, yet each of them provides some unique insights to the nature of decision-making processes in organizations.

Schön's study identifies the need for champions to shepherd new ideas through reluctant organizations. Champions identify strongly with the new ideas and promote them vigorously throughout the organization, using primarily informal channels. They are more successful if they are powerful figures who have a broad knowledge of how the organization works.

Maidique builds on the work of Schön and others by identifying the different roles involved in technological innovation and showing how they are a function of the size and stage of development of the organization. In the small firm, the entrepreneur performs virtually all roles. In the integrated firm, champions from middle management sponsor projects which emerge from the lower technical ranks; the entrepreneur acts as final arbiter. In diversified firms, executive champions provide another level of sponsorship by integrating the product champions with the entrepreneur or top management.

Bower provides a complex model of the factors that shape the resource allocation process. Rather than focusing on roles, he chooses to describe the activities performed by managers at different levels. Top managers provide a context in which lower level managers define proposals for new facilities; middle managers then provide impetus if the proposals look attractive. The attractiveness of a proposal has technical, business, and political dimensions. While final decisions are technically made by the board of directors, the real decisions are made at intermediate levels in the organization.

Carter's work illustrates how departmental loyalties shape the projects that are submitted by middle management for approval at higher levels. He notes that project submissions are often biased and presents variables that influence the extent of this bias. He also argues that the number of criteria used will change as a function of the uncertainty about a project. Finally, he reinforces the idea that a project sponsor's background will influence the way his or her proposals are received by upper management.

Burgelman's work is limited to a particular type of innovative activity: internal corporate venture development in a diversified firm. He portrays ICV development as consisting of definition and impetus activities similar to those described by Bower. But ICVs, in addition to being constrained by current strategy and structure, hold the potential for creating changes in both strategy and structure. Also noteworthy are the efforts of product champions to demonstrate the viability of new technologies, the lack of reliance on formal analytic methods for evaluating projects, and efforts by middle managers to develop the credibility on which this decision process places a premium.

Kanter distinguishes between two different types of organizations: integrative, which encourages innovation, and segmentalist, which discourages it. Individuals attempting to innovate in either structure will go through three waves of activity: problem definition, coalition building, and mobilizing and completion. The success of these activities will be determined by both individual management and interpersonal skills. More important, the type of cultural and structural context that is created by integrative or segmentalist thinking is crucial to this process.

Kanter gives more emphasis to this aspect of innovation than did any of the other works discussed. Perhaps the types of innovations that Kanter studied were more likely to have cross-departmental implications than, say, Bower's new facilities or Burgelman's new ventures. However, Carter did describe the process of negotiation employed among departments before investment proposals were sent to top management for approval, and Schön did note that champions needed to be able to work across organizational boundaries, although this was presented as an individual skill, with little emphasis on the organizational dynamics that help or hinder such efforts.

This selective review of the existing innovation process literature performs several functions for our study of AMT. First, it offers some guidance as to what issues, roles, and processes we should look for in the justification decision process. Second, it provides a context for understanding what is new and different—and what is *not* new and different—about the decisions associated with AMT. Third, since different types of innovations and projects were studied by the various authors, this review presents some preliminary conclusions as to how the decisionmaking process varies as a function of the nature of the project or proposal being considered.

REFERENCES

Bower, J.L. 1970. *Managing the Resource Allocation Process.* Boston: Harvard Business School Press.

Burgelman, R.A. 1983. "A Process Model of Internal Corporate Venturing in the Diversified Major Firm." *Administrative Science Quarterly* 28, no. 2 (June): 223–44.

Carter, E.E. 1971. "The Behavioral Theory of the Firm and Top-Level Corporate Decisions." *Administrative Science Quarterly* 16, no. 4 (December): 413–29.

Cyert, R.M., and J.G. March. 1963. *A Behavioral Theory of the Firm.* Englewood Cliffs, N.J.: Prentice-Hall.

Kanter, R.M. 1983. *The Change Masters: Innovation and Entrepreneurship in the American Corporation.* New York: Simon and Schuster.

Maidique, M.A. 1980. "Enterpreneurs, Champions, and Technological Innovation." *Sloan Management Review* 21, no. 2 (Winter): 59–76.

Pettigrew, A.M. 1973. *The Politics of Organizational Decision-Making.* London: Tavistock.

Rothwell, R.; C. Freeman; A. Horlsey; V.T.P. Jervis; A.B. Robertson; and J. Townsend. 1974. "SAPPHO Updated — Project SAPPHO Phase II." *Research Policy* 3: 258-91.

Schön, D.A. 1963. "Champions for Radical New Inventions." *Harvard Business Review* 41, no. 2 (March–April): 77-86.

Schwartz, J.S. 1973. "The Decision to Innovate." Ph.D. dissertation, Harvard University Graduate School of Business.

3 OVERVIEW OF THE STUDY

As Chapter 2 indicated, a number of excellent studies of the innovation process have been performed, ranging from novel human resource practices to newly developed product technologies. However, there has been little close examination of AMT justification. Some field research will thus be necessary to better understand the innovation decision process for AMT.

The objective of my study was to examine in depth the decision process by which investment in AMT is justified. What types of people are generally involved? What roles do they play? What are the organizational or other barriers to justification, and how are they overcome? What approaches do firms use in justification decisions? In this chapter I describe the study I undertook to answer these and other questions. I looked at AMT justification decisions in each of five different organizations, all of which were in different industries.

THE COMPANIES AND DECISIONS STUDIED

Due to the sensitive nature of this topic, the names of the companies and individuals I worked with have been changed here. This policy, which the participants were aware of, provided me much greater flexibility in discussing and reporting on the justification decisions.

International Metals Incorporated (IMI) is a large, vertically integrated producer of basic and engineered metal products. Through its various business units and subsidiaries, IMI operates facilities and serves markets around the world. I studied IMI's justification process for computer-integrated manufacturing (CIM). The CIM proponents originally wanted to devise an architecture for CIM that could be used throughout the corporation with any necessary customization. The decision I examined was not whether a particular CIM application would be utilized in a particular plant, but what broad changes would need to take place in the corporation to allow such specific decisions to be made. One of the most important of such changes was the attempt by technologists to gain the respect of influential marketing and financial managers. The IMI case is presented in Chapter 4.

Defense Technology Incorporated (Deftech) is, as the name indicates, a high-technology supplier of products used primarily in defense applications. (It is a subsidiary of a corporation that focuses primarily on defense-oriented products.) Deftech sells both directly to the military and to other defense contractors, who then integrate Deftech's products into their own systems. This study centered on Deftech's justification of a computer-aided design (CAD) system which would be used in both design and manufacturing areas. One notable aspect of the process at Deftech was its decision to justify CAD without performing any of the traditional financial analyses. Another was the attempt by the design and manufacturing departments within Deftech to coordinate their investigation and justification for CAD. The Deftech case is presented in Chapter 5.

Temple Laboratories is an old, established firm that offers a diverse line of products for both consumer and industrial markets. Many of these products are based on discoveries made by Temple's formidable cadre of scientists. I studied a plant that was involved in the consumer electronics business and was experiencing substantial cost pressure from foreign competition. In order to cut labor costs, Temple was considering a robotic material handling system that would replace a number of operators. This case illustrates the difficulties of implementing new technology at a time when the business itself is battling for survival, as well as the impact of unforeseen personnel changes at high levels on the justification process. The Temple case also provides some lessons about the advantages and disadvantages of an "islands" approach to automation. This case is presented in Chapter 6.

American Plumbing Fixtures (APF) is an old manufacturing company owned by a diversified corporation. APF produces plumbing fixtures for both residential and institutional use, and, like Temple, was in an extremely difficult competitive situation. The plant that I studied was outdated, very costly to operate, and experiencing difficulty with both the Occupational Safety and Health Administration (OSHA) and the Environmental Protection Agency (EPA). APF's management was faced with the choice either to invest in technology or to shut the plant down. They decided to invest in a revolutionary new process that utilized robotics. In order to take advantage of this process, APF had to transfer the technology from Japan. This decision was complicated by the financial, rather than technological, orientation of APF's corporate parent. The APF case is presented in Chapter 7.

The fifth and final company I studied is Monumental Building Supply (MBS). MBS was founded several decades ago by a prototypical entrepreneur, literally working in his garage. Its products are primarily aluminum windows, storm doors, and extruded aluminum shapes. MBS was bought out some time ago by a fairly large corporation that wanted to diversify, although its core businesses were in another industry. My study centered on MBS's efforts to justify a manufacturing resources planning (MRP II) system. This project was undertaken in an attempt to regulate a manufacturing environment that had gotten out of control through product proliferation and customer demands for frequent small deliveries. An interesting facet of this decision to innovate was that MBS had been forced by its parent corporation to coordinate its investigation and justification of MRP II with another subsidiary in the same product division. Thus there are some lessons about what is involved in coordinating justification between two separate and autonomous organizations. The MBS case is presented in Chapter 8.

The companies, industries, and technologies represented in these five cases are quite diverse. While I did not make a systematic attempt to create a particular mix of companies, the group appears to be a reasonable cross-section of American manufacturing. They represent consumer, industrial, and military markets. Some focus on a particular type of market and product, while others serve a variety of markets with a particular material. Their annual sales range from well below $100 million to much more than $1 billion. The scope of implementation for the technology considered by these firms ranges from one

Table 3–1.   Overview of Companies Studied.

Company	Industry	Sales	Technology	Focus
IMI	Metals	Over $1B	CIM	Corporation
Deftech	Defense	$50–100M	CAD	Company
Temple	Consumer electronics	$50–100M	Robotics	Plant
APF	Plumbingware	Under $50M	Robotics	Plant
MBS	Building products	$50–100M	MRP II	Company

product line in a single plant to a corporate-wide effort. The technologies under investigation vary from relatively hardware-oriented robotic systems to software-intensive technologies such as CIM. Information about these companies is summarized in Table 3-1.

THE STUDY

I began the study by contacting an individual in each firm who was centrally involved in the decision process I wanted to examine. In exchange for having participated in the study, each firm was promised an assessment of its own approach to justification, as well as the opportunity to compare its approach with those of the other firms participating in the research.

Once my initial contact had agreed to participate and had secured permission to do so, I conducted an initial (usually very long) interview. In this first meeting, I gathered background information on the firm, the industry, and so on. In the course of the interview, I obtained the names of other individuals who were involved in the justification process. I also interviewed these individuals, continuing the process until all the key participants were interviewed. The number of people interviewed in each firm ranged from four to eight.

The interviews were straightforward: I simply asked each participant to tell me his or her story of how the decision in question was being made. Most of them were both very willing and able to trace the steps and events involved in the process. Additional questions served primarily to follow up something a participant had said, to ask for additional information, or to clarify an uncertain point.

This "open-ended" method is intended to encourage the participants to describe the justification decision process using their own words and concepts, thus allowing the structure of the decision to emerge from their descriptions rather than from my expectations. It also leaves open the possibility of discovering aspects of the process that could not be anticipated from what is currently known about innovation decisions.

With the exception of Deftech, all of the decisions were studied in real time, that is, while they were occurring. The initial interviews focused on the history of the decision process up to my point of entry. Subsequent interviews dealt with events that had taken place since my last visit and included follow-up questions about earlier events. The total number of interviews conducted per firm varied from five to twenty-nine. Figure 3-1 illustrates the duration of each decisionmaking process, as well as the timing of my study in each firm.

Whenever possible, I also collected archival materials pertaining to each justification decision. While the amount and quality of this information varied from company to company, it typically included internal memos, minutes of meetings, letters to and from vendors, and documents presenting the financial justification for the technologies

Figure 3-1. Justification Decision Process Time Lines.

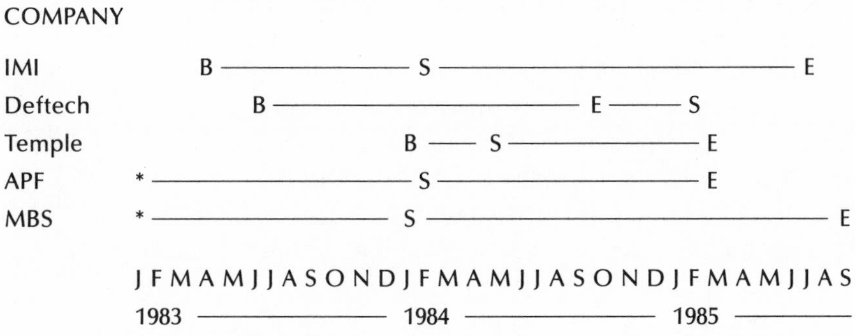

COMPANY

IMI	B ——————— S ——————————————— E
Deftech	B ——————————— E ——— S
Temple	B —— S ————————— E
APF	* ————————————— S ————————— E
MBS	* ————————————— S ————————————————— E

J F M A M J J A S O N D J F M A M J J A S O N D J F M A M J J A S
1983 ——————— 1984 ——————— 1985 ————

B = the beginning of the decision process.
S = the date at which the study was begun.
E = the end of the decision process.
* = the beginning of the decision process in two companies can be traced back as far as 1981.

being considered. This material proved quite helpful, especially in establishing the dates of various events and identifying additional people who had been involved in the decision but had not been mentioned by the interviewees.

The methods I used in this study are similar to those used by the researchers whose work was discussed in Chapter 2. Most of them relied, as I did, on semi-structured interviews which were augmented by review of available archival materials. While Bower (1970), Carter (1971), and Burgelman (1983) all studied one firm exclusively, Kanter (1983), Schön (1963), and Maidique (1980) looked at a number of firms in different industries. All of these studies attempted to uncover patterns in complex and apparently messy processes. With the possible exception of Carter, the emphasis was on generating new ideas rather than in testing predictions or hypotheses. Even Carter, who was ostensibly testing Cyert and March's (1963) theory of the firm, devoted most of his article to a discussion of his unexpected findings.

A natural question regarding this method is the extent to which the descriptions of the process were biased by the participants' desire to present themselves and their companies in the most favorable light. I attempted to avoid this possibility through a number of means. The first involved the real-time nature of the study. Because I was present for much of the process, there was little opportunity for the participants to rewrite history after the process was over. Also, the participants began to trust me because I was no longer just a one-hour interviewer with an assurance of confidentiality.

My second defense against biased answers was the open-ended nature of the interviews themselves. By asking a manager, "How was that decision made?" I was less likely to elicit bias than if I had asked, "Did you undertake a thorough analysis before making that decision?" In addition, the participants were more interested in explaining "how justification really works" than in creating the impression that the process conformed to any normative model.

The third protection I had against bias resulted from using several participants in each company. A major misrepresentation of a decision process would have required a complex and persistent conspiracy among participants; this would be extremely unlikely in an organization that had volunteered to participate in the study.

Fourth, the guarantee of anonymity limited any desire the participants may have had to make their companies look good in print. And finally, the archival material provided an additional check against the participants' portrayals of the decisionmaking process. Given all of this, I think it is highly unlikely that these representations of the justification process are materially wrong.

The interviews were taped whenever possible; they were then transcribed verbatim. (Taping was not feasible for interviews that were held over the telephone or in noisy restaurants. Overall, forty-four out of fifty-six interviews were taped.) The resulting transcripts, along with notes from untaped interviews and archival materials, comprise the "data base" for the study. (An indication of the volume of data collected is the 500-plus, single-spaced typed pages containing the interview transcripts.)

THE CASES

When the decisionmaking process in each firm had ended, and I had done as much as possible to gather information about it, I wrote a preliminary version of the case for the firm. The case was then circulated among the participants to assess its accuracy and completeness, and to provide them the opportunity to make suggestions for revisions. The case was revised following these suggestions; this process was repeated until everyone was satisfied that the case was accurate.

Disguising the case was my next task. This involved a trade-off in terms of the amount of information concealed or disguised. The greater the degree of disguise, the safer the identity of the firm—but at the cost of perhaps not providing some detail that would aid in understanding the case. In general, only the names of the companies, organizational subunits such as divisions or groups, individuals, and locations were changed. The firms varied, however, in the degree of disguise they required in order to release their cases. In some instances, the exact nature of the product, or the process technology being considered, was withheld. The description of the justification decision process itself was not changed or disguised in any way.

REFERENCES

Bower, J.L. 1970. *Managing the Resource Allocation Process.* Boston: Harvard Business School Press.

Burgelman, R.A. 1983. "A Process Model of Internal Corporate Venturing in the Diversified Major Firm." *Administrative Science Quarterly* 28, no. 2 (June): 223–44.

Carter, E.E. 1971. "The Behavioral Theory of the Firm and Top-Level Corporate Decisions." *Administrative Science Quarterly* 16, no. 4 (December): 413–29.

Cyert, R.M., and J.G. March. 1963. *A Behavioral Theory of the Firm.* Englewood Cliffs, N.J.: Prentice-Hall.

Kanter, R.M. 1983. *The Change Masters: Innovation and Entrepreneurship in the American Corporation.* New York: Simon and Schuster.

Maidique, M.A. 1980. "Entrepreneurs, Champions, and Technological Innovation." *Sloan Management Review* 21, no. 2 (Winter): 59–76.

Schön, D.A. 1963. "Champions for Radical New Inventions." *Harvard Business Review* 41, no. 2 (March–April): 77–86.

II THE CASES

4 INTERNATIONAL METALS INCORPORATED
The CIM Decision Process

This story of the advent of CIM at International Metals Incorporated is based on interviews with seven people who were involved in the decision process. This case, like the others, does not evaluate the actions taken by any of the participants; rather it describes what happened as accurately as possible.

THE RESURGENCE OF TECHNOLOGY

IMI is a large, mature corporation with several business units, all of which involve the metals business. A partial organization chart for IMI is shown in Figure 4-1; a list of participants in this study is given in Table 4-1. In the late 1970s, a number of people within the corporation began to feel that technology was not being taken seriously by the corporation. While this idea was expressed in different ways by different people, it centered around the perception that the company was overly concerned with finance and marketing and that these disciplines had replaced technology as the driving force within the firm. "Technologists" were not occupying key vice presidential positions, their statements were discounted; technologically risky decisions were not being made. In the period between 1978 and 1982, a number of efforts to combat this trend arose spontaneously in disparate parts of the corporation.

Figure 4–1. Partial Organizational Chart—International Metals Incorporated.

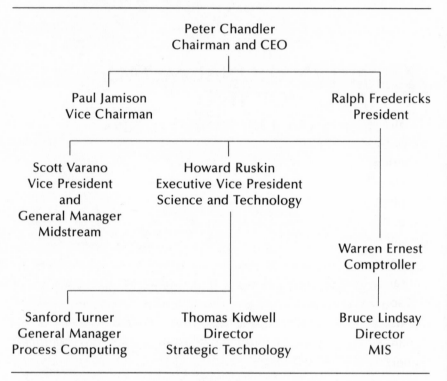

Steven Robinson, manager at the Corporate Research Center (CRC), had become increasingly frustrated with the R&D planning of the business units. R&D projects were funded by the business units, whose general managers also chaired the committees that allocated funds to research projects. Over time, it became clear that long-range, fundamental, "blue sky" projects were not getting funded. Instead, the business units were funding research that would generate quick returns on existing product lines. Robinson and others became concerned about the long-term effects of this type of planning on the corporation. In order to address the problem, an R&D planning group was formed at CRC. This group was to assist the committees in developing long-range R&D plans. It experienced some success in getting the business units to put research in a more long-term and strategic perspective.

At about the same time, Thomas Kidwell was transferred from CRC into corporate planning. He soon became concerned about the

Table 4–1. Important Participants—International Metals Incorporated.

Broadbent, Patrick	Process control expert at Corporate Research Center, affiliated with the Upstream business unit
Chandler, Peter	Chairman and CEO of IMI, formerly president
Charles, Nate	Member of Strategic Technology Group, former member of R&D planning group
Ernest, Warren	Comptroller of IMI
Evans, Helen	Member of Strategic Technology Group
Fredericks, Ralph	President of IMI
Jamison, Paul	Vice Chairman of IMI
Joseph, Tony	Member of Process Computing Group
Joyce, Donald	Assistant Director of the Corporate Research Center
Kidwell, Thomas	Director of Strategic Technology Group
King, Michael	Chief Electrical Engineer for Upstream business unit
Lindsay, Bruce	Director of Management Information Systems
Munson, Geoffrey	Former Vice Chairman of IMI
Robinson, Steven	Manager at the Corporate Research Center
Ruskin, Howard	Executive Vice President of Science and Technology
Turner, Sanford	General Manager of Process Computing Group, former Chief Electrical Engineer
Varano, Scott	Vice President and General Manager, Midstream business unit

harmful effects of the strategic planning process on technology. His concern is expressed as follows:

> Corporations, not just International, but corporations as a whole are getting increasingly into technological and business problems because there's an inordinate financial emphasis in the strategic planning process. Strategy is more than finance. Strategy expresses itself ultimately in finance, but it does not capture technology unless it asks for a moment, "What could technology substantively do for us?" You do that in the language of technology, and then, you can fold it into a financial plan. But if you don't do that carefully, you'll miss it.

Kidwell was unsuccessful in convincing top management of the importance of this problem, at least during this time period. He was hampered by his place within the financial organization and the fact that he had had three different bosses in his two years in corporate planning. "Each time I got some awareness in the vice president," he lamented, "he moved and I got a new boss." Thus Kidwell's message barely got beyond his department.

The third, and probably the most significant, initiative in the resurgence of technology at IMI began on a Sunday afternoon in 1982. Geoffrey Munson, who had one year to go before retirement as vice chairman of IMI, was having dinner at a local country club. He noticed at a nearby table a man who was a former vice president for a steel company. He said, "You know, it's beginning to haunt me that those men, instead of being able to enjoy this country club, should be punished for the way they've destroyed an industry." He wondered if the management at IMI also could be guilty of this. At that moment, he vowed that he would not be in a position to be accused of the same thing after he retired.

As a result of this experience, Munson began trying to incorporate technology into IMI's business plans by creating a new entity within the corporation: the Strategic Technology Group. Thomas Kidwell was named its director and reported to Munson. As Kidwell describes it:

> Munson knew he had to make it so that technology could rise to the top of the corporation without passing through finance. That meant that corporate planning and technology planning had to be parallel. I could not report to corporate planning. Technology had to rise to the top of the corporation in parallel.

The R&D planning group that Robinson had started at CRC now reported to Kidwell; its scope was broadened to include the whole corporation. The main task of the Strategic Technology Group was to assist the business units in preparing their plans to include technology. Nate Charles, who had worked in R&D planning and then in strategic technology, described the impact of the latter group:

> It's reached a point that in the forthcoming planning cycle, when the businesses bring in their annual five-year plans...they have been instructed by the president that they must deal explicitly with technology... [they must] give explicit examples of what their targets are, how much it's going to be worth once they reach those targets, and how they are going to get there. No broad-brushed stuff any more. No more glib words like "we're going to put in robots"... he said that you've got to be much more explicit than that.

Shortly after the formation of the Strategic Technology Group, IMI's office of the chairman changed *en masse*. The former chairman retired and was replaced as chairman and CEO by the former president, Peter Chandler. Geoffrey Munson also retired and was replaced by Howard Ruskin, who assumed roughly similar duties under the title of executive vice president of science and technology. The president's office was assumed by Ralph Fredericks. Paul Jamison remained as IMI's vice chairman; finance reported to him.

Howard Ruskin had the following comment on the situation at that time:

> I would suspect that the technical effort wasn't getting the proper coordination or push; it would be pretty easy for whatever came up to be washed away. [When the top level changes] is the best time for change. . . . It is certainly a time in which you can make changes that the top guys won't block. Whether the rest of the organization will block it or not depends on how good the top guys are. Also, it's a time that creates an awful lot of turmoil. Folks are jockeying for position. . . . Everyone is nervous.

Many aspects of Ruskin's sentiment were borne out by subsequent developments at IMI.

With the creation of the Strategic Technology Group, and the turnover of the senior officers, the seeds that Munson and others had planted began to grow and bear fruit. First, the new Executive Committee drafted a "Statement of Direction" for IMI. Both corporate planning and strategic technology were involved in the drafting of this statement — a significant development at IMI. The document also provided some direction for technology: "We will strengthen our core by focusing our resources where IMI can improve our competitive advantage . . . enter new areas and expand existing businesses that build upon our strengths. . . . In all our endeavors we will continue to be an innovative technological leader."

While such assertions can perhaps be dismissed as boilerplate, this was taken as a serious statement about IMI's commitment to technology and competition. As Bruce Lindsay, the director of management information systems (MIS) put it:

> I think there's a total commitment at the policy level of this company to have a rallying point. . . . We are going to plant a flag, we're not going to abdicate our business to foreign competition, and we're not going to sit and atrophy the way steel did. Here's where we stand and fight. That

implies we're going to do things differently, because [what steel did] was a formula for disaster. One school of thought would be that basic industry is a lousy business...we should get out of it, let the Third World have it, and start building semiconductors or something. We've said no, we'll change, this is our business. It's going to be our business ten years from now. Instead of looking for an easy solution, we'll do what we've got to do, to be here ten years from now, and hand off a healthy company to somebody else.

Subsequent to the executive turnover and the Statement of Direction, the resurgence of technology exhibited a number of tangible outcomes. The budget for CRC was increased, along with the expectation that it would not be the first thing to be cut back in the event of an economic downturn. A science advisory group, consisting of top people in various fields, was formed; its reports would be presented to the board of directors. Another new group, the Technology Council, was formed with Howard Ruskin as its chairman. This group includes the director of CRC, the vice president of engineering, the vice presidents of the business units, and the director of Strategic Technology. Its mission is to oversee the total portfolio of technology activity.

Thus there were a number of outcomes resulting from the resurgence of technology at IMI. One of them, the development of CIM, is the focus of the rest of this case.

THE APRIL 1983 MEETING

The Strategic Technology Group was formed in fall 1982. As we have seen, its primary role was to assist the business in developing a technological context and strategy as an integral part of their plans. In addition to coordinating this activity, Thomas Kidwell had taken on another task. With input from the executive level, he had developed a list of words or terms, each of which denoted a technological option that might be open to IMI. As he started his new job as director of strategic technology, he began to examine the list. The word that quickly rose to the top of the list was "computer."

Kidwell had not had a great deal of prior involvement with computers and was immediately fascinated by them. While most of the technologies he was exploring had price/performance ratios that were growing at a yearly rate of 2 or 3 percent, computers were growing at 25 to 100 percent—with no limit in sight.

In order to explore this technology, Kidwell and Helen Evans, a member of his group, spent the early months of 1983 visiting other firms. Many of the firms they visited, such as IBM, Digital Equipment Corporation, and General Electric, were both producers and users of computer equipment. Kidwell and Evans noted quickly that the emphasis on computers in these firms was quite different than that at IMI. While IMI was concentrating on business applications (e.g., accounting, payroll), other companies were involved with manufacturing applications (e.g., process control) and the integration of manufacturing and business systems. Kidwell and Evans were quite struck by this discrepancy.

In order to further IMI's progress in computer technology (which Kidwell had dubbed "low-cost information management"), he decided to have a meeting of those people within the firm who were most involved with computers — MIS, which is in the financial organization, the Process Computing Group, which at this time reported to the chief electrical engineer, and CRC, which reported to Ruskin. The meeting was planned for April 1983.

Kidwell gathered data on what he considered to be the wisest use of low-cost information. This data fell into two areas: the technology itself and the organizational arrangements necessary to support it. On this latter point, he noted that, for example, IMI would need a higher ratio of engineers to accountants. Kidwell then began what was to become a long and frustrating campaign to keep consideration of computer technology from being overwhelmed by organizational or "turf" issues. It soon became clear that computers and, particularly, computer integration were intricately entangled with the notion of what organizational arrangements would be needed to support this technology. Kidwell was directed explicitly to exclude discussion of such arrangements from the April meeting. As he put it:

> They told me, "How can you even bring up the subject? Don't you know how sensitive it is in the organization?" So this part got killed off. Organization was illegal. . . . We narrowed the subject area to say that even though we've learned some things about how the human system responds to computerization, that's not going to be dealt with in those three days at all.

As the meeting approached, Kidwell and Evans decided that its attendance should be limited to high-level people in the three computer-related areas. Bruce Lindsay, who had recently become director of

MIS and had an extensive background in operations, represented his group. Sanford Turner, the chief electrical engineer, represented process computing. Donald Joyce, an assistant director at CRC, represented the computer-oriented research part of the company. Each of the three was accompanied by two others from his area. Finally, two people from the business units were present; Kidwell and Evans participated as facilitators.

Many people in the company were amazed that the meeting could be held at all; MIS and the Process Computing Group had a long history of ignoring one another. The two divisions were located in two different buildings which were separated by a rarely crossed river. As Kidwell put it:

> Our human system began as completely separated. Four hundred and fifty MIS people degreed in business, accounting, and computer science. One hundred fifty [in process computing] with electrical engineering degrees, and no mobility between them. Nobody from here ever went there, and nobody from there ever came here. They don't even know who each other is. . . . They've never met each other before. It's a human problem.

The meeting was held on April 11–13, 1983. As promised, the guidelines distributed to the participants required that they "deal with what IMI should do with computers, not what organization best enables us to do it." The participants spent the first one-and-one-half days listening to presentations made by representatives from IBM, Digital Equipment Corporation, Arthur D. Little, and General Electric.

The representatives were asked to assess the need for computers at a manufacturing company like IMI. Kidwell and Evans devised a mechanism whereby 100 points were distributed by the speakers among the various potential computer applications: business computing, office automation, CAD, process computing, and so on. These point allocations were to express the representatives' suggested priorities for future computer projects within IMI. They did this to show the participants the gap they had perceived in their visits to the representatives' companies, that is, the discrepancy between IMI's (de facto) computer strategy and state-of-the-art computing. All of the representatives stressed the need for IMI to emphasize process computing and computer integration, that is, CIM.

The consultants' presentations were followed by a description of IMI's current deployment of computer resources. This presentation was made by Helen Evans, who had spent several months performing an audit of the company's use of computers. It completed the picture that began to form with the consultants' assertions: while they had advised that IMI emphasize process computing and integration, IMI was placing two-thirds of its resources in business computing, one-third in manufacturing/process computing, and virtually no resources in the integration of the two.

Faced with this information, the participants then were asked to enumerate potential computer strategies for IMI over the next decade and to arrive at a consensus on which strategies were most appropriate for the corporation. Consensus emerged around three strategies: (1) use computers as a tool to differentiate IMI from the competition, (2) increase computer literacy within IMI, and (3) utilize CIM; the third strategy rose quickly to the top of the list. The participants went beyond mere endorsement of CIM to advocating the immediate selection of a demonstration site:

Formulate and implement an integrated information system for IMI;
Select a location and implement immediately the integrated system
(including CIM) at a plant or business unit of manageable size to
enhance feasibility and benefits; and
Recommend that the upcoming modernization embody state-of-the-
art CIM.

Thus in just a three-day meeting, the top computer professionals at IMI came to an agreement on computer strategies for the next decade and how they could be initially implemented. The next major step would be to obtain Executive Committee approval for these strategies.

What actually happened at the April meeting? First of all, the participants apparently *did not* learn anything new about CIM. As Bruce Lindsay put it, "Tom felt more strongly than the rest of us about the consultants being there. They didn't say anything that anyone who reads *Datamation* wouldn't know...a lot of hype, no real insights." Sanford Turner also noted that everyone pretty much knew what the experts were going to say. Lindsay, Turner, and Tony Joseph (a member of Turner's group) all had some previous interest or experience with CIM.

If the meeting was not primarily an educational experience for the participants, why was it universally seen as important? Several things were apparently accomplished. Sanford Turner felt that there were big contributions from the meeting—the participants were buffered from their pressing day-to-day concerns, and a direction was established:

In an operating entity, the problem is today's business, and that's where you gravitate all the time. You have to get off and think about new and innovative things, which is very difficult. So we were looking for approval of a direction that would allow us to go off and worry about CIM [and] keep ourselves out of the mainstream of daily problems.

In addition, while the consultants' ideas were not seen as a big revelation, the fact that they were in agreement with one another did seem to have an impact. As Lindsay put it, "What they lent was a catalyst. Four different perspectives, all with a common theme, without rehearsing,...made people feel a little bit better.... Any one of a dozen [internal people] could have gotten up and said it, but I don't think it would have carried the same weight."

Finally, in spite of — or perhaps because of — the proscription against discussing organizational arrangements, some barriers were overcome. Bruce Lindsay commented:

I think Tom really threw a spotlight on the need to work together. You've got four computing communities in the company: the research group, the business computing group (MIS), the process computing group, and the plants. To get anything done, those four end up having to coordinate and work together. The overlap had been minimal, and the interfacing had been only when necessary. The thing that's getting increasingly apparent to everyone is that...we've all got a vested interest in [CIM], let's work on it together.

Perhaps the real significance of the April meeting can best be captured by an exchange that took place between Helen Evans and Thomas Kidwell when the meeting was over. Evans told Kidwell that she was disappointed in the outcome of the meeting because the strategies that had been adopted were obvious before the meeting took place. Kidwell responded:

Wait a minute, there's a difference between you having made up your mind [on the strategies] as a result of spending four months outside, and the corporation forming consensus and commitment around this word "computers." In a sense, the meeting was to develop the backdrop, the common commitments, and the working relationships to do something about it.

Evans, after reflecting on this, concluded: "Out of everything that happened out of this meeting, this was the most valuable thing. These guys closed ranks. They put away politics for a while, and said, "Hey, there's a technology out there that we ought to be grasping." As subsequent events would confirm, the closed ranks, common commitments, and working relationships would be absolutely necessary in surviving the challenges that lay ahead.

THE EXECUTIVE COMMITTEE

Following the April meeting, the team of Sanford Turner and Bruce Lindsay, with the help of Thomas Kidwell, tried to schedule a presentation before the Executive Committee. They were seeking top-level corporate support for the computer strategies. However, the secretary of the committee did not consider this an appropriate agenda item for the committee; thus the presentation was made to only a few committee members, as well as to other key individuals in IMI's senior management.

The presentation was not very well received. Descriptions of the reactions to it included "a bloody nightmare," "great abuse," "blown out of the water," and "thrown out." In fact, this abuse was not restricted to the presentation itself. As one of the presenters mentioned, "we were beat up all week long."

The participants were unanimous in concluding that, once again, technology had been defeated by organization. During the time of the meeting, IMI was undergoing a major push toward decentralization and business unit autonomy. The executive officers thought that Lindsay, Turner, and Kidwell were advocating recentralization through the use of computers. According to Ruskin, Kidwell, and Turner, problems were inherent in the timing, the wording, and even the identity of the presenters. Their comments follow:

> The first thing that happened was that it was a political thing. It was not a technological argument, it was an organizational argument, and this kind of slowed things down a lot. It was turf: who was going to do what.... I think that just clouded things.... At the time, we were doing all of this reorganization...and the word "organization" came up, and it kind of worried them. (Ruskin)

> The subject of centralization/decentralization had gotten confused with the subject of networking and architecture. Those of us who had considered computerization had never dealt with the question of whether or not computers should be used in a centralized or decentralized corporation.... You can draw the corporate lines either way.... But when we were talking about words like "architecture" and "networking" and so

on, people thought that meant centralization at a time when they were trying to be decentralized. (Kidwell)

At the time we did this, we were in the throes of decentralizing [and] they wanted no inference whatsoever that the integration effort was going to centralize [the corporation]. They had just committed their souls and a lot of people's livelihood [to decentralization], and they couldn't segregate the two. Maybe if we would have had a representative from the two or three major business units with us as presenters, the officers would not have read what they read into it. But here were Bruce and I . . . both of us corporate, making this presentation on computer strategy. . . . It came across that these strategies were going to be corporate mandated. (Turner)

The computer spokesmen did not back down in the face of this reaction, a fact that did not go unnoticed by the senior management. Turner, Lindsay, and Kidwell believed there was nothing substantive in what they had heard; they would not change their minds and they concluded that the problem was "basically semantics." Thus they tried to recover from the presentation debacle by using a number of related tacks.

First, anything in the strategies that even hinted at organizational issues was deleted. Second, they removed any language that sounded, even remotely, like computer jargon. They were concerned that the uncertainty created by this language may have been threatening to the officers, none of whom were experienced in computing. Third, the strategies were "softened" a little, by emphasizing consultation with the business units, rather than direct action.

In order to defuse any further misinterpretation of what they had in mind, Kidwell, Turner, and Lindsay held a number of one-on-one meetings over a two-week period with the senior officers. Kidwell met with Ruskin, his boss, as well as with Jamison and Fredericks. Turner met with Chandler, at his request. Lindsay talked with his boss, Comptroller Warren Ernest. The officers were eventually convinced that this was not an organizational issue — at least in the minds of Kidwell, Turner, and Lindsay. Another meeting was scheduled for the officers to discuss the strategies with their proponents. This resulted in a set of revised strategies that were explicit in their recognition of business unit autonomy:

Emphasize the technological importance of the key elements in business unit computer strategies including CIM; and

Work with the business unit managers to select a location of manageable size, and begin to immediately implement a computer integrated manufacturing system.

The wording changes and explanations that were used in the one-on-one meetings had the combined effect of swaying the officers; at the second meeting, the strategies were approved. This approval, however, was accompanied by an even further commitment of the officers to decentralization. The computer spokesmen were responsible for the sale of their strategies to the business units. "Get your story together," they were told, "and then go talk to the businesses. If they support it, then we support it."

To emphasize the importance of the CIM initiative, the Process Computing Group was elevated one step in the corporation. Rather than reporting to the chief electrical engineer, it now reported to the vice president of engineering. Sanford Turner was chosen to head the group and, to underline the move, was given the title of general manager.

THE BUSINESS UNITS: ISSUES

Having obtained corporate consensus on broad strategies for computer utilization, the computer spokesmen would now have to convince the vice presidents and general managers of IMI's three major business units to spend money on CIM. The next hurdle would be to secure funding for demonstration sites; this presented a new set of issues, including the length of time it would take to see any return, the intangibility of the short-term products of the investment, and the large amounts of money that would need to be spent initially. As Bruce Lindsay put it:

> What's just alien to a lot of management is that we [want to spend] at one location a million and a half dollars to do nothing but a general design. That still hasn't dawned on the operating folks. They expect to spend a million and a half dollars and get a 50 percent ROI next year. . . . It's just how we are conditioned to go at things. I think that's one of the more fragile dimensions of the whole process. Some folks are going to spend a million and a half dollars, and what they are going to get is four thousand sheets of paper that tell them they have a real bear to take on, versus a product.

Then, you say to carry it through the detail designs, it's going to take ten years and $20 million or whatever it is. When you get into those kind of numbers, rather than being a nicety in the corner, it is going to be center stage, because now you are starting to compete with major capital. There's going to be a lot of people saying, "Wait a minute. When I buy a cold mill, I know what I get. When I buy this, all I know is I got a bunch of computer types running around saying this is the right thing to do. I didn't grow up with it, I don't understand it, and I'm not really sure what the hell I'm going to get out of it." I think we still have to cross that bridge before we are really off and running.

Another issue involves the criteria that would be used in evaluating investments in CIM. Due to intense competition for corporate capital, IMI's hurdle rate for cost reduction projects (which would include computer projects) had been increased from 50 percent to 100 percent. It became clear that it would be very difficult for CIM to meet these hurdle rates; thus there was something of a mismatch between the computer strategy and the finance policy. IMI's technological resurgence perhaps contributed to the resolution of this issue:

These projects have to be justified to some extent, but there's a little bit of the rigor removed from the intensity of the justification.... Once the corporation begins to lean toward CIM, then the individual proposals have a better chance of passing the guidelines. (Kidwell)

[Years ago] the technical people said we don't really need computers. You can't justify them.... They held back the development of the computer for a long time. In these kinds of technologies, you have to get ahead, and do some degree of testing it out. Get it out of the conversational stage, get a critical mass in there, move it.... There's hardly any way you can IE it or MBA it to really find out. You've just got to try it. (Ruskin)

I'm financial and quantitative and analytical by nature, and I don't think that [ROI] is the right question to be asking. I think the question is, "Do you want to be in the business?" What does it take to be successful at that? If this is one of the things that it takes, I don't think you have an option not to do it.... Our comptrollers today have a very heavy [operations] kind of background. They use accounting as...one of many tools.... It's usually false precision when you start reducing things to columns and rows anyway. They'll run through that to make sure they're in the general ballpark, but within that I think they agonize with the general manager about the market and the risk, and the technology.... Ultimately, I don't see [the numbers] driving our decisions. I think that's very healthy. (Lindsay)

A third issue that had to be addressed before CIM implementation could proceed was, inevitably, organization. Put simply, who would do the work? IMI's reorganization had included a substantial downsizing of corporate engineering. A large number of engineers were relocated to the plants or given early retirements; some had even quit. Thus a lack of engineering resources made corporate handling of the projects difficult. However, many believed that the plants did not possess the resources to implement CIM either. Even if they did, it would be difficult to create a truly "integrated" system from a set of autonomous, plant-level groups. If corporate were to develop an architecture at one plant that would then be used at others, who would pay for it? Perhaps the most daunting organizational issue was that CIM had to be "created" at IMI through some combination of technical support from Electrical Engineering, MIS, and CRC—a coalition that had simply never before existed in the corporation. All of this, complicated by the pervading spirit of decentralization, put CIM's future in doubt.

In the second half of 1983, Turner, Lindsay, and Kidwell began to approach the business units with their proposition. In spite of the difficulties to be overcome, Lindsay was optimistic about the business units' response to CIM. He felt that what was most necessary was an initial success.

> I would guarantee that the business and plant managers, *if you demonstrate to them the ability to meaningfully change their ability to compete,* would go through the hammers of hell with you. Their staying power is greater than that of a functional person when they believe there is something at the end of the road. (Emphasis added).

THE BUSINESS UNITS' RESPONSE

The Northwestern plant, which is part of the Midstream business group, had been mentioned by name at the April meeting. Those attending that meeting had felt that the size of the plant was appropriate for a demonstration project; there also was some interest in CIM among the technologists there.

In July 1984, Scott Varano, vice president and general manager for Midstream, approved $900,000 to be spent on a requirements definition for CIM. This work was to be done by Sanford Turner's group, with the help of some MIS personnel, and was to be completed by late 1985 or early 1986. To date, the personnel at Northwestern have demonstrated good support and enthusiasm. Some have credited Tony Joseph, a member of Turner's group who is responsible for Northwestern, with solidifying process computing's relationship with that plant's production management.

The Downstream business unit has displayed a more problematic response. When CIM emerged from the April meeting and the Executive Committee as an area of emphasis for IMI, Downstream was already in the midst of a complete modernization effort involving three plants. The Midwestern plant was chosen over the Southern plant as a demonstration site for CIM because it was smaller and had a greater volume of technical skill.

Once the personnel at Midwestern began to talk to Sanford Turner about how the job would be done, however, tensions arose. They had always seen themselves as an independent plant and felt that Turner was telling them that it had to be done his way. A falling out occurred between Midwestern and the Process Computing Group. The type of bond that had been formed between the corporate group and Northwestern never materialized here, and the relationship came to be characterized by "a fair degree of animosity."

As a result, Midwestern decided that, rather than utilizing corporate services, they would hire an outside consultant. The consulting firm had never done a CIM job outside their own facilities, and most corporate computer personnel were convinced that their approach to the Midwestern project was much too narrow. This effort is still in process at Midwestern.

The third and final business unit that showed interest in CIM was Upstream. Following the April meeting, Kidwell, Patrick Broadbent (from CRC), and Michael King (chief electric engineer for the business unit) held a similar meeting in October exclusively for personnel from the business unit. Some of the same outside presenters used at the April meeting were present. Broadbent and King both had some prior interest in CIM; Broadbent had participated in the earlier meeting.

As a result of these efforts, the Upstream group approved $350,000 for the development of limited CIM capacity for the foreign plant. This project also was to be undertaken by an outside group. Turner and others were disappointed with the scope of the foreign effort. However, it was clear that the corporate group did not have the resources to pursue this project. Ironically, the foreign initiative in CIM had to be discontinued altogether. A world oversupply of its product in early 1985 forced the foreign plant to shut down.

PRELIMINARY ANALYSIS

From both a technical and organizational standpoint, IMI's project was the most ambitious of any of the cases I studied. CIM is currently the state-of-the-art in manufacturing technology; only a few firms have been able to succeed with CIM in one plant, let alone corporate-wide. IMI's progress toward a CIM architecture was marked by a number of events, some of which helped their cause, and others that impeded it.

The CIM initiative could never have happened without Geoffrey Munson's experience at the country club. His prediction that IMI would end up just like companies in the steel industry unless technology was taken seriously was the catalyst for the creation of the Strategic Technology Group. While Munson was never involved directly with CIM per se, organizational changes he instituted were a necessary condition for CIM to flourish.

CIM was given its first opportunity at the April meeting. This meeting was significant because it was the first attempt at cooperation between MIS and process computing. (However, it almost didn't take place because of the corporation's concern that recommendations about organizational arrangements, rather than just technology, would come out of it.) The meeting produced, among other things, a consensus that IMI should be actively pursuing CIM, as well as attempting to identify some demonstration sites.

The first presentation made by Turner, Lindsay, and Kidwell was probably the lowest point in the entire effort. The presenters were not successful in separating the technical issues of networking and architecture from the organizational issues of centralization. In fairness to both sides, however, these issues are really *not* separate; it is an open question as to whether the presenters could have conceivably avoided this problem at a time when the senior management group was unwilling to see the presentation in any other way. Of course, the fact that all of the presenters were from the corporate staff did not help to avoid the perception that they were trying to shift the balance of power away from the business units to the corporate level.

The CIM advocates were able to recover from this debacle only through a combination of explanation and persuasion that required a considerable display of resolve on their part. Most people at IMI probably expected them to give up on CIM at this point. It was when they did not give up, but instead redoubled the efforts of their fragile coalition, that people began to take them seriously.

Having secured some degree of support from corporate management, the CIM advocates then went to the business units to try to get funding for some demonstration projects. They enjoyed a fair level of success in one of the three major business units. This particular project was progressing well at the time this case was completed.

In the second business unit, the spectre of organization came back to haunt the CIM advocates. While this unit did undertake a CIM initiative, it contracted an outside group rather than the IMI corporate staff group to implement it. This business unit wanted to make it clear that this was their project; they resented the apparent degree of influence that the corporate group was trying to exert. This decision was the final blow to the idea of a corporate-wide CIM network and represented the final triumph of turf over technology at IMI. Global competitive forces forced the third business unit's target plant to close, and CIM quickly became a nonissue.

POSTSCRIPT

In the early months of 1985, the Science Advisory Board, which had been formed by Howard Ruskin, began to look at IMI's computer strategies. Its report stressed the fragmented nature of the corporation's computer initiatives and the unevenness of the progress being made toward CIM. It described the company as "fragmented," "a set of baronies," and "lacking focus." Thomas Kidwell, who agreed substantially with the advisory board, said that the situation reminded him of a passage from scripture: "In those days there was no king in Israel; every man did that which was right in his own eyes." (Judges 21:25)

5 DEFENSE TECHNOLOGY, INCORPORATED
The CAD Decision Process

Defense Technology, Incorporated (Deftech) began many years ago in a consumer products industry. Over the years, particularly after World War II, consumer products at Deftech were gradually replaced by products for the military. This evolution culminated in the 1970s with the formation of Deftech's current corporate structure, whose primary mission became the development and production of defense products. The federal government is Deftech's major customer. In the past five years, Deftech's annual sales have grown from about $25 million to close to $100 million. Figure 5-1 displays a partial organizational chart for Deftech. A list of participants in this case study is included in Table 5-1.

THE INITIAL SEARCH

Three sources were responsible for Deftech's interest in computer-aided design (CAD), which surfaced around 1981. First, President and CEO Bruce Kennedy was encouraging the use of computers throughout the company. Second, the manufacturing area discovered CAD's potential for programming computer numerically controlled (CNC) machines. And third, the design area was looking for a way to automate the drafting process in designing products, tools, and equipment.

Figure 5–1.   Partial Organizational Chart—Defense
Technology Incorporated.

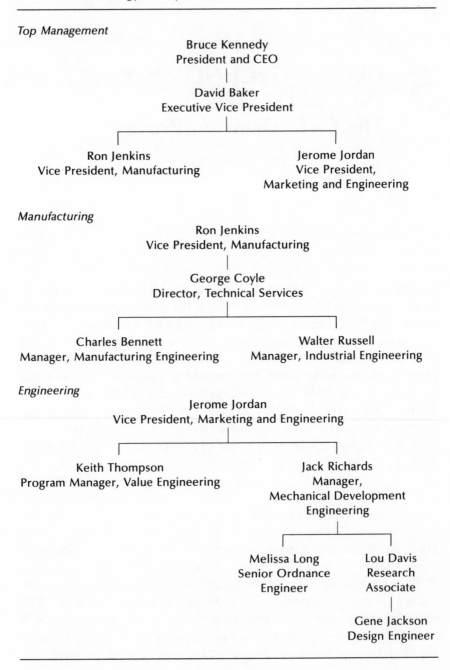

Top Management

Bruce Kennedy
President and CEO

David Baker
Executive Vice President

Ron Jenkins
Vice President, Manufacturing

Jerome Jordan
Vice President,
Marketing and Engineering

Manufacturing

Ron Jenkins
Vice President, Manufacturing

George Coyle
Director, Technical Services

Charles Bennett
Manager, Manufacturing Engineering

Walter Russell
Manager, Industrial Engineering

Engineering

Jerome Jordan
Vice President, Marketing and Engineering

Keith Thompson
Program Manager, Value Engineering

Jack Richards
Manager,
Mechanical Development
Engineering

Melissa Long
Senior Ordnance
Engineer

Lou Davis
Research
Associate

Gene Jackson
Design Engineer

Table 5–1. Important Participants—Defense Technology, Incorporated.

Baker, David   Executive Vice President
Bennett, Charles   Manager of Manufacturing Engineering
Coyle, George   Director of Technical Services
Jackson, Gene   Product Design Engineer
Jenkins, Ron   Vice President of Manufacturing
Jordan, Jerome   Vice President of Marketing and Engineering
Kennedy, Bruce   President and CEO of Defense Technology
Long, Melissa   Senior Ordnance Engineer
Russell, Walter   Manager of Industrial Engineering
Thompson, Keith   Program Manager for Value Engineering

Both design and manufacturing were spending small amounts of money on an investigation of CAD.

Initially, design personnel were more active in investigating CAD. In particular, Keith Thompson, program manager for value engineering, and his staff were involved. (Value engineering is a program by which firms attempt to reduce the cost of a product. If a firm is successful, it shares in the cost savings of other suppliers who were directed by the government to follow the design change, as well as benefiting from its own cost savings.) Design's approach to the investigation at this time (early 1982) was somewhat informal and consisted primarily of visits to nearby firms that had implemented CAD. Some manufacturing personnel also visited these user firms.

In late 1982, George Coyle was promoted to director of technical services and began pushing CAD in manufacturing. At about the same time, Ron Jenkins became vice president of manufacturing. Coyle asked Jenkins (his boss) to require manufacturing engineering (under Coyle) to conduct a bona fide study of CAD. Thus manufacturing began its own investigation. This took place independent of, and at a lower level of activity than, design's efforts. The two groups were, however, aware of each other's efforts.

By early 1983, Thompson's group at design had made several additional visits to CAD sites and had spoken with a number of vendors. They decided to push for the purchase of CAD equipment from Bausch and Lomb—a purchase that was resisted by George Coyle. He contested that the Bausch and Lomb system consisted primarily of automated drafting and did not include certain other features, notably networking, desired by people in manufacturing. The design group countered that this feature could be developed at a later date. Eventually, it was decided by Ron Jenkins in manufacturing and Jerome Jordan (vice president of marketing and engineering) that the selection of a CAD system would have to be a "joint venture" between manufacturing and engineering. By this time, however, relations between the two groups were somewhat strained.

Following the Bausch and Lomb episode, Deftech's president directed that the investigation of CAD be coordinated under Coyle. The idea of a true joint search was not realized, partially because the meetings between the two groups were becoming shouting matches. During this period, however, value engineering was able to ascertain what manufacturing wanted in a CAD system.

Coyle's next step was to enter into a contract with a local university, part of which included the undertaking of a study of Deftech's need for CAD. Coyle felt that a systematic study of the problem would provide for input from all relevant parts of the company and would ultimately help them to determine which system would best fit their needs. Thompson objected to this plan on the grounds that the university researchers had never stated clearly what the product of the investigation would be and that their limited CAD budget should be spent on hardware, not study. The study became the second major bone of contention between design and manufacturing. However, it was approved, but at a lower level of expenditure than was conceived originally.

Within six months, the university investigation was behind schedule. Coyle grew disappointed with its progress. His disappointment was paralleled by design's disgust. Design insisted that what they needed was automated drafting, not "the factory of the future" (the broader scope of the university study). While manufacturing shared information obtained from the study with design, there were, from design's point of view, "no results."

RENEWED SEARCH

In late 1983, approximately a year after the university study was commissioned, Keith Thompson was "elected" by design to do something about electronic drafting. A CAD committee consisting of Thompson, Melissa Long (a senior ordnance engineer), and Gene Jackson (a product design engineer) was formed. The committee met throughout the end of 1983 and into 1984, engaging in an intensive search for the best CAD system for Deftech. Thompson, Long, and Jackson gathered information from some twenty-five vendors, went to a number of trade shows, and visited additional CAD users. At least ten CAD sales personnel also made calls to Deftech during this period.

By this time, Thompson's group saw themselves as representatives of the entire company in their search for CAD. Coyle had familiarized them with manufacturing's requirements, and they kept these in mind during their search. However, relations between the two groups had not yet improved to the point where the CAD investigation could be considered a joint project. Thompson gave periodic reports on his progress to his boss, Jerome Jordan, who was very concerned that Thompson's efforts not exacerbate tensions between the groups.

People from quality also were briefly involved in the investigation; they were interested in using the CAD plotter to make inspection charts. After some study, Thompson decided that a CAD plotter would not be precise enough to generate inspection charts. This set of requirements thus was dropped from consideration.

Since no team could research all of the vendors, price and performance criteria were used to limit the detailed investigation to a manageable number of companies. Systems from Hewlett-Packard, Lockheed, McDonnell-Douglas, and Auto-trol were reviewed in detail. Thompson's group decided to move toward a three-dimensional CAD system which would provide greater analytical capability than two-dimensional systems like Bausch and Lomb's. The search was eventually narrowed to McDonnell-Douglas and Auto-trol. McDonnell-Douglas had a mainframe-type system, while Auto-trol's system consisted of networked workstations, each with its own processing capacity. This system had been used previously by an individual in tool design. After more than eight months of investigation, the group decided to recommend the Auto-trol system.

By this time, money for a CAD system had been placed in the approved 1984 capital budget. Coyle, Thompson, and others also had begun to sell the idea that the investment in CAD would be unusual in that there would be no normal documented payback. The benefits would be somewhat intangible and therefore would not lend themselves to traditional financial justification techniques. By early 1984, top management was demonstrating some level of acceptance of this approach.

In the summer of 1984, Thompson approached Coyle with the news that his group had decided on Auto-trol as the system that would best fit the company's overall needs. A meeting was scheduled so that Thompson's group could share its research and conclusions with manufacturing. Coyle's reaction was extremely positive. He described the project as "a complete, excellent job of research" and complimented the design group on their efforts. Plans then were made to share this presentation with the president and his staff.

Manufacturing then began their own investigation of Auto-trol. Coyle, Charles Bennett (manager of manufacturing engineering), and Walter Russell (manager of industrial and plant engineering) visited the Auto-trol plant and user sites. Sales personnel from Auto-trol again visited Deftech, this time to talk with manufacturing. After their investigation, Coyle and his group came to agree that Auto-trol was the system they wanted.

THE PRESENTATION TO MANAGEMENT

In August, Thompson met with the president's staff. The president; Executive Vice President David Baker; vice presidents of manufacturing, program management, and finance; and MIS director all were present. According to Coyle:

> We don't usually [have this kind of presentation] on these justifications. We usually route them, and they read it and they sign or ask questions. This was different in that it involved a lot more people, and there was no documented payback. It was strictly technological, staying abreast of what we needed to do, with no apparent payback. In fact, it would cost us money for a while, until we learned how to use the thing. It was a matter of survival in the fast-moving world of electronic computerized communications.

At this meeting, a basic agreement was reached: this was a very viable program. The next step was to obtain cost estimates and prepare the authorization request. As Ron Jenkins, vice president of manufacturing, later recalled, "I think that meeting was probably the turning point. It was pretty obvious at that point that we had sold the right people that this was the thing we wanted to do.... Everyone said 'This is for real; I think we can do it'."

During August and September, activity continued on two fronts: vendor negotiations and preparation of the capital request. At first, discussions with Auto-trol were held separately by the design and manufacturing groups. Eventually, however, the purchasing department also got involved by coordinating communications and negotiations with the vendor. This worked to the company's advantage; purchasing was able to extract a discount from Auto-trol, contingent on the purchase of two workstations each in fiscal years 1984 and 1985.

Efforts were simultaneously underway to prepare the justification package for senior management. Coyle, Thompson, and Charles Bennett, who coordinated this effort, were in agreement that traditional calculations of ROI were inappropriate for CAD. As Thompson put it,

> We could have attempted to go to a study where you could come up with numbers, and you could say, your payback is this. But, because there were so many estimates, it becomes nonsense. There's no point in trying to do that. You can make of it whatever you want with an estimate. Whatever assumptions you make, you come out with a new answer. So, it ended up that this was a business-type decision where you have a gut feel that "yes, this is what we are going to do."

Bennett was even more succinct in his appraisal of the situation: "I couldn't come up with a way to say that in two years, we'll be operating for 'this much less' because we are spending $300,000."

Coyle and Bennett thus solicited statements from the four immediate user groups — product design, drafting and documentation, tool and equipment design, and industrial engineering — testifying to the advantages of CAD. Coyle encouraged the users to focus on CAD's benefits and future potential, not on its paybacks. Since an individual from tool design had previous experience with the Auto-trol system, his input was solicited first and was shown as a model to the others. To ensure that they were not overly technical and that their advantages would be readily apparent to managers who were not involved in the day-to-day operations of the user departments, the statements were revised several times.

By the beginning of October, both the vendor quotes and statements of advantages were complete; the authorization request was ready to be presented to the executive staff. On October 9, a draft authorization order was circulated to the staff so that they would have time to read it before attending the next week's meeting. As George Coyle said, "We didn't route it in the normal manner. We routed a draft and called it that. We presented it to the staff after they had had a chance to read it. We wanted to do everything we could to pave the way for this program."

As advertised, the authorization order (AO) did not show a calculation of financial return; that part of the form was left blank. In its place were sixteen pages of narrative that detailed the advantages of the proposed system for each user group. Many of the advantages fell into the categories of improved/increased capabilities and routine task time savings. Particular mention was made of the fact that CAD would allow Deftech to participate in "electronic bidding" with the government. An overview of the justification was included with the package:

> Traditional cost justification techniques are inappropriate for this type of purchase. Benefits are generally long range, and do not lend themselves to a "present vs. proposed" method of analysis. . . . As outlined above, CAD will allow us to do what we do better and faster, and allow us to do some things we presently cannot do. We are dealing with quality and content of output as much or more than we are dealing with quantity. The true economic benefit of CAD is the avoidance of presently unplanned future costs.

After the staff reviewed the proposal, a formal meeting of all relevant personnel was held on October 15. In order to set the tone for the meeting, Coyle pointed out that there were only thirty-eight slide rules sold in the last year — and they were all sold to museums. (He later admitted that he had no idea if this was true.) In contrast, he continued, there were millions of calculators sold during the same period. He concluded that the choice of CAD versus drafting boards was the same as that between a calculator and a slide rule. The meeting then progressed to a formal presentation of the justification for CAD.

Although Bruce Kennedy asked a number of questions concerning both cost savings and equipment selection, the meeting was relatively smooth sailing for the CAD proponents. Charles Bennett expressed some surprise that the meeting wasn't any rougher:

> I thought we'd really have a problem with the finance people, but I think they saw the fact that it looked like everyone else really wanted and needed

this. Bruce really seemed to be going for it.... You couldn't really argue with the numbers, because there weren't any numbers to argue with. Maybe we would have had a harder time if we had tried to show a payback.

Executive Vice President David Baker called an additional meeting for the majority of participants later that same day. It was again relatively smooth for the CAD proponents. The only new issue raised was whether to buy two or four workstations. A compromise was eventually reached: two stations would be bought immediately, two more would be purchased next year.

The only issue that remained unresolved was which vice president would sign the AO. Ron Jenkins suggested that since this had been a joint project, both he and Jerome Jordan, the vice president of marketing and engineering, should sign it. David Baker did not approve, suggesting that because manufacturing would ultimately be the biggest user, it would have to sign and take responsibility for making the system work. Jenkins ended up signing the approval form.

By October 19, all of the required signatures had been secured. The system was approved. When George Coyle went to tell Keith Thompson the good news, he found Thompson talking with a couple of people from design. When one of these designers heard the news, tears of joy came to his eyes.

How was it that Deftech was able to spend $300,000, which represented approximately 25 percent of its annual capital budget, on an investment with no calculated return? Ron Jenkins provides one perspective:

The support of top management had been there initially for this kind of system. I don't think this was a case where the users had to do a heck of a big sales job. I don't know how you explain that environment.... Whatever the reason,...although this company is very frugal and generally looks at things in dollars and cents,...we also are blessed with a management that says, "We're here to grow and do things for the future. If we don't stay involved with the way the whole world is going, we're just going to fall by the wayside."

For Deftech, to invest that kind of capital, it's a rather significant thing.... We did go through a rather detailed presentation of what this could do for us over the long haul, because the immediate tangible results of the system are not very apparent.... But we looked at the problem areas that the company had to address, and [CAD] was a good way to do it.

Another important consideration, according to Jenkins, was the relationship between design and manufacturing:

> Another intangible, but very important, consideration is to get the proper interface between Design and Manufacturing. That's a problem in any company, and we're no different from anyone else. Just the fact that we had a common goal right here was an important step forward. You have two different groups saying, "This is the kind of system that will work for both of us and bring us together." I think that was a big consideration in the minds of a lot of us.

Charles Bennett added an additional perspective on their success in getting the system approved:

> I think that being enthusiastically behind it, being a booster, is what does it. We are the people our vice president relies on. So, when we get behind something and are very confident about it and when we do our homework, it's tough arguing it.... It's not sold in the meetings. It's not sold with the executive vice president sitting there reading it. It's sold by getting met in the hall...and saying, "We'll really need that CAD, absolutely, we've got to do it." That's the sort of thing that sells this thing; the meeting just formalizes it.... I really believe you get a lot of this stuff by being 100 percent positive in the informal...contacts.

Coyle later calculated a memo congratulating all involved personnel (see Figure 5-2). He noted the vendor evaluation efforts of Thompson's team, as well as the foresight shown by the executive staff in approving CAD. As Coyle later commented, "we wanted our management to know we're proud of them."

PRELIMINARY ANALYSIS

The Deftech case provides a number of insights. The first, which echoes certain aspects of the IMI case, is the difficulty of coordinating the efforts of two different groups in the justification process. The conflict between the two groups was a significant impediment to the CAD effort. The resolution of the conflict was probably hastened by the fact that both groups recognized that they needed CAD, that they were both functional groups supporting the same business unit, and that they were located in the same building.

Second, Deftech is certainly unusual in that it approved the CAD system without a traditional calculation of return. In fact, I have not

Figure 5–2. Congratulatory Memorandum.

Defense Technology, Inc.
Director of Technical Services
October 1, 1984

To: Distribution
Subject: CAD Systems

As a follow-up to receiving authorization and issuing a purchase order for our Computer Graphics systems, I would like to congratulate all involved personnel for taking this very important step to enter into this extremely important technology.

First, all who participated in the long and detailed systems investigation are to be commended. Keith Thompson and his team performed a distinct service in reviewing many hardware and software vendors. Also, a parallel effort was ongoing in the Technical Services group. The ranks were then closed and we came to a common decision to recommend the Auto-trol systems.

We are particularly pleased that our Executive Staff had the foresight to approve this technological advancement, allowing us to get in step with the industry leaders as well as ready ourselves for the coming electronic information transfer era. This is a sure sign of a mature management philosophy—one that will allow us to continue our technological improvements, maintain our upward growth and secure our future.

George L. Coyle
Director of Technical Services

---

heard of any other cases where AMT was approved in this manner. The CAD proponents' decision to try and sell CAD as such was no doubt influenced by their belief in the futility of calculating returns using such uncertain estimates.

Third, the CAD proponents' success can in part be attributed to the physical proximity that brought the two groups together in the first place. Thompson, Bennett, Coyle, and others could informally sell the CAD system to the senior officers without having to make a special trip to do so, or even appearing to be doing so intentionally. (Bennett made a point of mentioning the importance of accidental meetings in the hall.)

Fourth, Coyle and Bennett's efforts to "translate" the expected technical benefits of CAD into business objectives that could be appreciated by senior management also played a key role. While Bruce Kennedy's intimate knowledge of operations culled from years of experience in the business would have made it easier for CAD to be sold on technical criteria alone, it would have been more difficult to sell it to the financial managers without these efforts at translation.

Finally, the Deftech case stresses the importance of being willing to take responsibility for making the system work. This is illustrated by the question that arose when it came time to sign the authorization request. While it may have made sense for both vice presidents to sign it, Executive Vice President Baker wanted to make sure that he would know which one to look for if the system was not being used as promised. As the chosen signator, Jenkins was having his confidence in the proposed system tested. As Baker put it, if Jenkins was not willing to take responsibility for making the system work, he would not get it.

POSTSCRIPT

Deftech's implementation of CAD has been successful to date. The current plan is to increase the total number of CAD workstations to nine, and the original four workstations are being updated. Through the implementation of CAD (as well as other advanced technology projects), Deftech has been able to quadruple its sales volume in the past few years while only doubling its work force.

6 TEMPLE LABORATORIES
The Robotics Decision Process

Temple Laboratories' plant located in Hanover produces components for consumer electronics products. It functions twenty-four hours a day, seven days a week, producing thousands of units in a single day. Approximately 750 people are employed at the plant, in which the blue-collar personnel are unionized. Labor represents a substantial portion of product costs at Hanover. Temple has one major domestic competitor.

Consumer electronics (CE) components is one of two product lines in the Electronics Division at Temple. (This division is, in turn, one of three that comprise the Consumer Products Group. There are two other groups within Temple: the Scientific Group and the Communications Group. (See Figure 6–1 for a partial organization chart.)) Although the CE business represented two-thirds of Temple's earnings at one time, it is less profitable today. The reason for this is that Temple's customers have been losing market share to their Japanese competitors. This trend gradually led to the closing of all of Temple's CE plants except Hanover.

Competitive pressures have led to a strong emphasis on cost reduction at Temple. Each plant is given an annual goal for reducing actual manufacturing cost (AMC). For example, Hanover's 1984 goal was to reduct AMC by 6 percent. In order to reach these goals, Hanover embarked during the eighties on a number of automation projects that

Figure 6–1. Partial Organizational Chart—Temple Laboratories.

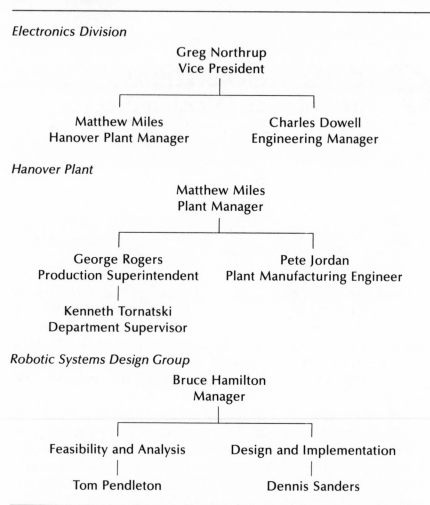

Electronics Division

Greg Northrup
Vice President

Matthew Miles
Hanover Plant Manager

Charles Dowell
Engineering Manager

Hanover Plant

Matthew Miles
Plant Manager

George Rogers
Production Superintendent

Pete Jordan
Plant Manufacturing Engineer

Kenneth Tornatski
Department Supervisor

Robotic Systems Design Group

Bruce Hamilton
Manager

Feasibility and Analysis

Design and Implementation

Tom Pendleton

Dennis Sanders

have reduced the labor component of product cost. Most of these projects consist of "hard" automation, and many of the displaced personnel are absorbed through attrition. This case will focus on the most ambitious automation project attempted to date at Hanover, the Material Handling Mechanization (MHM) project. (Participants in this case are included in Table 6–1.)

Table 6–1.   Important Participants—Temple Laboratories.

Dowell, Charles   Engineering Manager, Electronics Division

Hamilton, Bruce   Robotics Manager, Engineering and Manufacturing (E&M)

Jacobs, Harold   Machine Technology Manager, E&M

Jameson, Ronald   Project Manager, E&M, formerly Hanover Plant Manufacturing Engineer

Jordan, Pete   Hanover Plant Manufacturing Engineer, formerly Equipment Engineering Supervisor

Miles, Matthew   Hanover Plant Manager

Minert, Carl   Senior Industrial Engineer, E&M

Northrup, Greg   Vice President, Electronics Division

Pendleton, Tom   Member of Feasibility and Analysis Group, E&M

Sanders, Dennis   Member of Design and Implementation Group, E&M

Tornatski, Ken   Department Supervisor, Hanover Plant

THE ENGINEERING AND MANUFACTURING DECISION

In 1983, when one of Temple's CE plants was closed, Hanover acquired some of its technology. This technology was new to Hanover, which had previously utilized a different technology to make the same product. When they learned that the line was to be installed at Hanover, plant personnel suggested that it be put into operation as an automated line. More than a dozen employees involved in material handling, a repetitive task in a hot environment, would be eliminated as a result of this automation.

This suggestion was made to the "board of directors" for the move, consisting of Charles Dowell, the electronics division engineering manager, Walter Jeffries, the facilities equipment engineering services manager, and Matthew Miles, the Hanover plant manager. This group decided that it would be too difficult to move and mechanize the line at the same time; the request thus was turned down. Harold Jacobs, however, supported mechanizing the line after start-up difficulties had been overcome. He promised to commit E&M resources to do the project when the time arrived.

Hanover adapted very quickly to the new line. The line was installed in November 1983, and by December, Equipment Engineering Supervisor Pete Jordan had received permission from Plant Manufacturing Engineer Ronald Jameson to pursue the MHM project. Shortly thereafter, Jameson was promoted to a project manager position within E&M, and Jordan moved up to plant manufacturing engineer.

Jordan contacted Carl Minert, a senior industrial engineer in the Systems Design Group, to initiate the project. (Systems design is part of Machine and Manufacturing Systems, which is, in turn, part of E&M. E&M is part of Temple's corporate staff and performs engineering services across the corporation. Systems design is comprised of specialists in hard automation applications.) Minert and others from his group toured Hanover's facilities as a first step toward designing automation options. The next step was a careful documentation of exactly what was needed. For example, Jordan had requested that there be no material handlers on the system, and that it operate within additional cycle time, temperature, and space constraints.

Based on these ground rules, Minert's team began to identify design options for the system. One possibility that emerged immediately was to recreate the type of mechanization developed by a European firm with which Temple had a technical agreement. This option involved putting the machines on carousels, and would have cost between $5 and $6 million.

The E&M engineers developed four other designs, all of which employed carousels. Two of these options involved putting the existing machines on carousels—one utilizing one carousel, another two. The other two options required new machines. Again, there were both one and two carousel options—the "double dedicated" option would cost over $4 million, with a projected IRR of 16–18 percent.

A number of potential problems were anticipated with the carousel arrangements. First, the maintenance on the machines would have to be performed by a technician working on a moving carousel. Second, "rotary unions" would have to be created for the delivery of gas and oxygen to the machines. Third, the carousels would be quite large and would not fit easily within the existing layout of the Hanover plant. It was felt that the carousel options were somewhat risky, particularly because developing this technology would be a new venture for E&M.

Another option developed early on was the workcenter concept. In this arrangement, four workcenters would be created, each containing two of each type of machine. One operator would be stationed at each workcenter, where he or she would load and unload the machines.

Carl Minert had responsibility for quantifying the cost reduction to be derived from each design. When these preliminary calculations (plus or minus 25 percent) had been made, the various options were reviewed by higher level managers within E&M. The managers felt that none of the designs were justifiable, that is, they did not generate a large enough return. (Temple's hurdle rate for cost reduction projects is 20 percent.)

At about the same time, several individuals from the Robotics System Design Group got involved with the MHM project. The robotics group is also within the Machine and Manufacturing systems part of E&M. Its charter is actually broader than the name would indicate and includes automated guided vehicles and CIM. The robotics group consists of three subunits: feasibility and analysis, controls and sensors, and design and implementation.

Within a few weeks, Tom Pendleton (from feasibility and analysis) and Dennis Sanders (from design and implementation) and others from robotics, designed a preliminary robotics solution. The robotics option kept the machines stationary on the floor and grouped three machines around each robot. The robotics engineers pointed out that robots were proven technology and had been implemented in other Temple plants. The only design work required would be the end-of-arm tooling. The robotics option also had financial advantages. At about $3 million, it was the least costly of the five designs. It would generate over $1 million in cost savings annually, leading to a rate of return of over 30 percent.

In spite of these apparent advantages, systems design did not immediately reach consensus on the robotics concept. The group wanted to go to the plant with one voice, but, as one participant in the process described it, "we argued for days over the rankings of the options." The manufacturing engineers were very interested in designing one of their original solutions. After having spent a great deal of time on the other alternatives, they were at least a little reluctant to embrace an essentially ready-made robotics solution. Although the group decided to present all of the alternatives to the plant, they recommended robotics.

THE PLANT DECISION

On March 7, 1984, Carl Minert presented the various options to the Hanover plant. He shared the evaluation process of the E&M group and its matrix of weights and rankings, as well as its conclusion that robotics was the best option.

Reactions to the presentation were varied. Pete Jordan knew as soon as he heard the presentation that he wanted robotics, noting both its lower cost and flexibility. Others were not so sure. Robots had previously been used at Hanover, with mixed results that had led to their removal. The safety of the operators who would work near the robots and the reliability of the machines themselves were concerns. More recent successful employment of robotics in the plant had improved people's perceptions to the level of ambivalence.

Two additional plant concerns were addressed in a follow-up meeting between Minert and the plant, held the following week. First, the plant wanted the flexibility in the robotics option to increase the production rate if necessary. This was resolved by adding another workcell to the design. A second concern had to do with the ability of the design to match parts as they emerged from the cell. The E&M group had set up a static simulation to demonstrate that this could indeed be accomplished.

Shortly after the second meeting with Minert, the user group met one evening to consider the options. This group consisted of Ken Tornatski, department supervisor; Ben Flathers, section foreman; Keith Brazelton, a technician; Donald Paulus, equipment engineer; and Mike Johnson, a senior process engineer who worked with the new line. The group decided to evaluate each option individually in terms of performance potential, disregarding its calculated financial return. The "single dedicated" option was dismissed immediately on the grounds that if the carousel were to go down, production would be greatly curtailed. Although the group was leaning strongly toward the "double dedicated" carousel option, it considered each in its turn.

Robotics was initially greeted with hisses and jeers from the group. Beyond the safety and reliability concerns noted above, people were worried about the scrap rate problem. When a piece broke, a human operator could remove it from the machines, but a robot would continue trying to place subsequent pieces, thus breaking them and multiplying the damage incurred. Robotics' chief positive attribute was its flexibility. While redesign of the machines might render some hard automation designs obsolete, this would not be the case with robotics. The group felt that the robotics option would be adaptable to future, unforeseeable production arrangements.

When all options had been discussed, a vote was taken by secret ballot. Consensus was reached — all five individuals had voted for robotics. Pete Jordan had been away at the time this meeting was held. On his return, he was happy to learn that the user group had met and had endorsed the use of robotics on the new line. By May, Jordan told E&M that the plant had agreed on the robotics concept. E&M could then begin work on a number of fronts. First, the robotics concept now could be designed in much greater detail and could be tested using a mock-up of the system. Second, a process improvement unit within E&M could begin to devise potential solutions to the scrap rate problem.

The submission of the authorization request for the project was delayed until some progress was made on the scrap problem. (Engineering work was being funded by expense money that did not require an approved request.) By July, the process improvement group from E&M had made some advances on the problem and presented preliminary suggestions to the plant. They continued to work on the problem for the remainder of 1984.

THE DIVISION DECISION

The funding plan for MHM called for E&M and Hanover to split the cost of the project. The robotics group was charged with getting automation into Temple; it received some funds to facilitate the process. The group believed that MHM was a very worthwhile project and was willing to pay for half of it.

In late summer 1984, this arrangement began to unravel. Oliver Franklin, the division vice president, had recently left to take a job with another company, and Gregory Northrup, the new vice president, was keeping very tight control over capital for cost reduction projects. Half of the limited available capital was earmarked for introducing new products; some of the remainder had to be spent on legally mandated improvements. A number of attempts to free the capital for MHM were made in late summer and early fall. Pete Jordan made a presentation to Gregory Northrup on the need for cost reduction capital in general and the MHM project in particular. He pointed out that most of the smaller cost reduction projects ("nickel and dimers") had already been done. If Hanover was to make further progress in cost reduction, it would need substantial capital.

A number of issues emerged from the decision whether to allocate major capital to Hanover, the central one being the future of the domestic CE business. The declining market share of the American CE manufacturers, and its impact on Temple's business, has already been noted. Two of Temple's top executives were asked about the future of the business. The former president (currently the vice chairman) commented, "I'm very reluctant to say, even as difficult as the domestic CE business is, that we should pack it in." The chairman was less optimistic: "Something dramatically different has got to happen for us not to be out [of the business in the United States]."

In a sense, Temple was following the dictates of the business portfolio matrix approach. Hanover's profits were going to support growth in other businesses with higher potential. Such allegiance to the theory of corporate strategy, however, was cold comfort to those in the CE business.

Certain factors militated in favor of continued involvement in the business, one of them being the possibility of a high-growth U.S. market in display (computer) monitors. Temple was already supplying some parts for this market; it could experience substantial sales growth if the market took off. Another positive factor was Temple's relatively new business in CE component *factories,* which the company had started to sell overseas. In order to maintain credibility in that business, the base of expertise would have to be kept alive.

By early fall 1984, Hanover had received its capital. This was seen by many as a commitment to remain in the CE business for a long time. However, the sigh of relief that this produced at the plant was short-lived; in the interim, the arrangement between E&M and Hanover to split the cost of the project had disintegrated.

The problem was caused by a combination of poor timing and a misunderstanding between Pete Jordan (the Hanover plant manufacturing engineer) and Bruce Hamilton (the E&M robotics manager). When Jordan received word that capital would be allocated for the project, he asked E&M personnel to write the request. He was proceeding on the assumption that E&M would pay for half of the project. Hamilton notified him that, due to the delay caused by the capital allocation decision, E&M's 1984 money had been spent on other projects. Additional funds might be available in 1985 (when Jordan wanted to do the project), but this could not be guaranteed.

Jordan claimed that E&M never mentioned the need to spend the money in 1984. Hamilton claimed that it was his job to get automation into Temple as a whole; waiting a year to do a project would put him behind schedule. He noted that the funds designated originally for Hanover had been spent on a project with a higher return. He therefore felt his decision was justified.

After considerable discussion between Hanover and E&M, it became clear that E&M simply did not have the money to support the project. Jordan felt that he could not justify paying E&M's original share with plant funds. Combined with his own share, this would mean spending over one-half of the plant's $6 million capital budget on one project. Thus the project would not be done unless other funding became available. As of October 1984, the unsigned request languished on Jordan's desk.

This time the logjam was broken by the division engineering manager, who told the plants that he might be able to find an additional $1 million in capital. Both Hanover and another plant in its division were asked to submit proposals for cost reduction projects that could be funded by this capital. The proposals were due in early December.

In response to this request, Pete Jordan submitted four proposed projects on behalf of Hanover. The first was a small computer project for $100,000. The second was a machinery project that would cost $1.5 million (which was not to be spent entirely in the next year). The third was the material handling project, which had a lower projected IRR than the machinery project. While the total cost of the MHM project was about $3.5 million, Jordan would have been able to arrange it so that only $1 million would be spent in 1985. A vision system for detecting defects was fourth on the list. Because its feasibility was "sketchy," this was given the lowest priority.

When the proposals had been received, Gregory Northrup, the division vice president, called a meeting of his staff to discuss them. Matthew Miles, the Hanover plant manager, also attended. The vice president asked the division engineer and the controller to present ideas on how to free enough capital to do *all* of the projects. After some discussion, it was determined that this could be accomplished by reducing both raw materials and finished goods inventories. The division vice president decided to fund all the projects, including the MHM, and directed both the division staff and the plants to work on ways to reduce the amount of money tied up in inventory.

Soon after the division decision, a meeting took place between involved personnel from Hanover and E&M. They needed to rewrite the AR to reflect some changes that Hanover deemed necessary. As Jordan commented, "Our desire was to submit the $3 million Appropriations Request and get all the sign-offs [very quickly], because if you don't strike, and you delay them for three or four months, then something's going to come up to change their minds."

Unfortunately, E&M rewrote the request in 1984 rather than 1985 dollars. The Hanover plant comptroller directed that the financials be redone. Rather than delay the process further while this was being done, Jordan submitted the original request but marked it "reading copy only." A one-page cover request asking for a quarter-million dollars in capital and some expense money was attached so that E&M could begin building a prototype cell. By the end of 1984, the advance AR had all of the required plant and division approvals. Thus the E&M work still could be done in time to schedule the project to coincide with the 1986 planned shutdown — the biggest anticipated "window" for some time.

It would still be a matter of months before the MHM project received final approval. One minor setback occurred when the division comptroller wanted to change the time frame for making the decision on releasing the biggest portion of the funds. He wanted this decision point to be after the prototype cell had been proven effective. This request required reworking of the proposal as well as new approvals by the plant and the division engineering staff. Approval was secured from the division comptroller, Division Vice President Gregory Northrup, and corporate E&M before March. Final approval was granted by the board of directors in summer 1985; the prototype cell was installed that fall. As of October 1985, the installation of the MHM robotics was scheduled for late 1986 or early 1987.

PRELIMINARY ANALYSIS

The Temple Laboratories case is somewhat typical of that of American manufacturers encountering global competition in the late twentieth century. Temple was literally fighting for survival in the consumer electronics business, faced with very stiff competition from the Far East. In order to remain competitive, Temple's CE business would have to lower its costs; the robotic materials handling system under consideration was but one means to achieve this end.

In this situation, management is aware that it needs to invest in capital in order to survive. Ironically, however, corporate management is often not sufficiently sure that the business will exist long enough to invest in it. Although the materials handling project had a large, unambiguous payback based on labor cost reduction, the questions concerned whether the system would work, and whether the business was one in which the corporation wanted to continue.

The use of simulation to demonstrate the feasibility of the MHM system was a good move on the part of the robotics proponents. In addition to providing the answers to technical questions, simulation provides a way for skeptics to "see" that the system will work. This is particularly true for the animated simulations that are becoming more popular today. This type of visualization makes it much easier for senior managers to get a feel for the workings of a proposed system.

The Temple case also illustrates the fragile nature of the coalitions that often support efforts such as these. The Hanover plant and the robotics group were able to reach agreement temporarily on funding for the MHM project, but a delay in project approval caused this agreement to unravel. This difficulty in coordinating the work of corporate staff and plant personnel echoes some of the difficulties encountered by IMI. The turnover at the division vice presidential level also complicated and delayed the process of securing approval for the MHM project.

POSTSCRIPT

The MHM project was never implemented. The initial reason for this was that the material breakage problem was never resolved satisfactorily. While work was proceeding on this, the Hanover plant was designated one of Temple's CIM demonstration sites. In order to prepare for CIM, division personnel undertook a thorough examination of the processes used to manufacture each of the plant's products. The rationale for the study was that it would not make sense to automate and integrate a manufacturing process that needed improvement. This study resulted in a blueprint for the "process of the future," a process sufficiently different from the current one that the MHM system was no longer appropriate. Temple appears to have made a significant commitment to the CE business. As of early 1987, the plant was being expanded and CIM implementation was in the planning stages.

7 AMERICAN PLUMBING FIXTURES INCORPORATED
The Process X Decision Process

Since its foundation around the turn of the century, American Plumbing Fixtures Incorporated (APF) has steadily grown to become a major player in several plumbingware markets, including bathtubs, water closets, and sinks. It has four product lines: china, brass, cast iron, and fiberglass. APF's ownership has changed several times during its history. In 1954, it was acquired by Bennett Corporation, and in 1964, when Bennett merged with the Watson Company, APF became part of the Bennett-Watson Corporation. Finally, in 1982, Bennett-Watson was acquired by Diversified Corporation. After some initial reshuffling, APF became part of Diversified's manufacturing subsidiary. Diversified Corporation's 1983 sales were between $6 and $10 billion, of which the manufacturing subsidiary accounted for about 10 percent and APF for about 1 percent. (See Figure 7-1 for a partial diagram of Diversified Corporation's organization.)

APF LEARNS ABOUT PROCESS X

This case describes the process by which APF arrived at a major innovation: a new process for the manufacture of bathtubs and other cast iron plumbingware. (Key participants in this case are listed in Table 7-1.) Elmer Davis, manager of manufacturing engineering, has

Figure 7–1.   Partial Organizational Chart—Diversified
Corporation.

Clifford Yates
Chairman
Diversified Corporation
|
Dennis Gallagher
President and CEO
Manufacturing Subsidiary
Diversified Corporation
|
Wayne Balston
Group Vice President
Diversified Corporation
|
Dale Johnson
President
American Plumbing Fixtures
|
Fred Barnes
Vice President, Operations
American Plumbing Fixtures
|
Elmer Davis
Manager of Manufacturing Engineering
American Plumbing Fixtures

been with APF since 1948; he has been at corporate headquarters since
1969. One of his duties is to investigate new, state-of-the-art manufacturing techniques.

Davis attended the foundry show in Cleveland in 1973 where demon-
strations of a new casting method known as "Process X" were being
given. This process had been developed in Japan and was being intro-
duced in the United States at that time. Green sand molding, the most
common method used in foundries, involves complex sand chemis-
try and strict control over temperatures. In contrast, Process X uses
sheets of very thin plastic, a vacuum, and ordinary synthetic sand to
form the mold for the pieces to be cast. While the green sand "jolt

Table 7-1. Important Participants—American Plumbing
Fixtures Incorporated.

Balston, Wayne   Group Vice President, Diversified Corporation

Barnes, Fred   Vice President of Operations, American Plumbing
Fixtures Incorporated

Davis, Elmer   Manager of Manufacturing Engineering, American
Plumbing Fixtures Incorporated

Gallagher, Dennis   President and CEO, Manufacturing Subsidiary,
Diversified Corporation

Johnson, Dale   President, American Plumbing Fixtures Incorporated

Yates, Clifford   Chairman, Diversified Corporation

---

and squeeze" process necessitates very large, heavy-duty equipment, Process X affords automation with lighter duty equipment because the vacuum does the work.

When Davis saw the Process X demonstration, he considered the possibility of using it in APF's Hancock cast iron plant, where bathtubs and other products were being made with traditional green sand molding techniques. The representatives at the foundry show were making only small, simple castings. When Davis inquired about casting larger pieces with Process X, they replied that they were not at that time involved with anything very large. He thus dropped the idea.

THE RISE OF FIBERGLASS

At this time, only about half the companies remained of the dozen or so that had produced cast iron tubs. APF and two others emerged as the leading firms in the market. The exodus was due primarily to the diminishing market for cast iron tubs, which were rapidly being displaced by less expensive fiberglass models. Since fiberglass tub production requires a relatively small capital investment, many small firms had entered the market and were drawing business away from the established producers. The cast iron manufacturers were further hampered by problems in complying with OSHA and EPA requirements. Because of these developments, the other major firms had decided to enter the fiberglass market.

APF commissioned a leading management consulting firm to investigate the future potential for cast iron and fiberglass tubs. Their report expressed the belief that cast iron bathtubs were a thing of the past and recommended that APF enter the fiberglass market. APF's management took this recommendation seriously and devoted a great deal of attention to developing production capacity for fiberglass tubs. Between 1973 and 1980, no further substantial investment was made in the Hancock cast iron plant; Bennett-Watson's management decided that only maintenance activities would be conducted there.

Initially, APF contracted with a fiberglass tub manufacturer from the South; this manufacturer produced tubs with the APF name on them. By 1975, APF had designed its own tub and opened a plant in the Midwest to produce them. Shortly thereafter, Bennett-Watson purchased a fiberglass company and combined it with APF's Midwest plant to form Bennett-Watson Fiberglass, a separate division from APF, from which APF would purchase its fiberglass tubs. Because of labor problems, the Midwest plant was eventually returned to APF.

APF's management eventually determined that they could not compete in the fiberglass market with their existing cost structure. So, when the union at the Midwest plant refused to accept wage concessions, the plant was closed. In 1978, a new plant was opened in the Northeast to support the large fiberglass tub market there. However, the recession of 1981, along with the associated decline in housing starts, caused such a large drop in orders that this plant was also eventually closed.

THE PROBLEM AT HANCOCK

In the summer of 1980, Fred Barnes was promoted from his position as director of divisional planning and administration to vice president of operations. Before he took the job, Barnes had determined that something dramatic had to be done about the Hancock cast iron plant; this became his first priority almost immediately upon beginning his new job. At that point, the only improvements that had been made at Hancock over the last thirty-five years were those mandated by OSHA or EPA. Barnes described the plant as being in "pretty sad shape."

The first step in Barnes's initiative was a detailed evaluation of APF's two major competitors in the cast iron bathtub market. The costs associated with each of Hancock's processes were compared with those of the competition. The major conclusion of this evaluation was alarming: APF had the highest costs in the industry. A competitor who had installed an automatic bathtub machine in the early 1970s was the lowest cost producer.

Barnes's second step was to have an engineering study conducted, which would investigate the technological options available to APF at Hancock. Several contractors were screened, and Newton Engineering was eventually selected. Newton was instructed to consider the automatic machine, even though Barnes had serious doubts that such an investment could be financially justified. Newton was also asked to consider the possibility of using Process X.

The study was conducted in 1981 and 1982. Newton determined that it would now cost about $25 million to install an automated line similar to that which APF's competitor was using. Thus, as Barnes had suspected, such an investment could not be justified. Newton also advised against switching to Process X, believing that it was not sufficiently developed. Barnes was disappointed with the study; it did not inform him much beyond what he already knew.

At about the same time that the engineering study was being conducted, Elmer Davis attended a trade fair in Frankfurt, Germany. There he met some representatives of a French company that was making a long, luxury-model bathtub. Davis asked their sales representative to visit APF in order to explore the possibility of marketing these tubs in the United States. During the representative's visit, Davis learned that the French company was experimenting with making European-style "apronless" tubs using Process X.

Davis wanted to reinvestigate the possibility of using Process X for bathtubs, so he asked another acquaintance in the French firm if he and others from APF could visit to investigate the possibility of using Process X. The French company was very receptive and invited the APF personnel to make the trip. However, the late 1981 recession caused severe cutbacks at APF, and the trip had to be postponed.

APF became a part of Diversified Corporation at this time through Diversified's purchase of Bennett-Watson. After completing the cost evaluation and engineering studies, Fred Barnes was very concerned about the future of Hancock:

> Between all the research we did comparing ourselves to the competition, ...the fact that our profits were going straight downhill, and the fact that we had a tough union situation, a lot of things had to come together. My thought was that you either do something substantial or you really milk it. Don't spend nickel one and let the damn thing die a natural death.

But that product line ties in so closely with APF's success. . . . They are all separate plants, but they are all intermingled from a marketing viewpoint, and cast iron and china go hand in hand. So it was crucial for APF's success. . . . You couldn't look at it just as a single plant decision, it had to tie in with overall corporate strategy. I also knew that the plant was going to collapse. I had OSHA and EPA problems. I had high cost problems, due to aging equipment, high maintenance costs, and high product losses. . . . I had to either do something new and exciting or forget it.

Neither the automatic machine nor Process X seemed very promising. The automatic machine was technologically feasible but prohibitively expensive. With Process X, however, the problem was technological feasibility. In addition to Newton's recommendation against using Process X, it was not supported by the people at Hancock, some of whom had seen the process in operation in Europe. They argued that the process was five or ten years away from being ready to use in large-scale production. Barnes believed that if the people in the plant didn't think it would work, then it wouldn't work. An impasse was reached. As Barnes said, "I didn't know what the hell we were going to do if we didn't come up with something unique."

THE EXPLORATION OF PROCESS X

Two things happened to move the process forward. First, Barnes was able to secure major wage concessions from the union at Hancock in their 1982 negotiations. He believed it was necessary to change the wage structure before he could recommend major capital investment in the Hancock plant. Second, Elmer Davis had not abandoned the Process X exploration when the European trip was delayed. In late 1982, he discovered that Samson Molding Machine Company was now representing the Japanese firm that held the Process X patent. Davis was surprised that Samson had not contacted APF about Process X, especially since Samson had built the original foundry equipment at Hancock. In any event, he contacted Samson, and they immediately sent a representative. Together they began to develop the information that was needed to determine whether Process X would be viable at Hancock.

Through these explorations with Samson, Davis found that there was a firm in Yugoslavia that used Process X to make apronless bathtubs on a production basis. While the firm was currently only making about twenty tubs per hour, the process was rated at forty tubs per hour, about the same rate as the Hancock plant. According to Barnes, when Davis brought him a drawing of this line, "the light came on." Barnes spent a few minutes making calculations based on line speed, yield, and cost. If APF could get a line that would do a little better than the Yugoslavian line, he felt, "it just might be the answer to our problems." He decided to "go after this thing and see if it makes any sense at all, because it was our best bet as far as I could determine. That started the ball rolling."

Samson was then called in for extended discussions with Barnes and Davis. They soon discovered that a Japanese company was using Process X to make apronless tubs at a rated sixty molds an hour. The question remained whether Process X could be used to make American-style tubs with aprons; making the apron requires a very deep "draw," which would require heavy-duty automated equipment. As far as anyone at APF knew, American-style tub production had never been accomplished.

Because a trip to Japan would cost over $20,000, Barnes decided to start by visiting American firms that were using Process X. While none were making bathtubs with the process, they were making other products. Barnes, Davis, and a number of people from Hancock made several visits, concentrating on firms that made large pieces. However, at least one visit was made to a high-volume aluminum casting plant.

The most significant visit was to a facility in East Chicago that made tank parts. As Davis tells it:

> It was a 4,000-pound casting and it was in a flask about half the size of a room. They were pouring 8,000 pounds of steel when they made it, and the section through was about two inches thick. Now, they did not have an automated line, and they only made a few of those in a shift, but it was a tremendous accomplishment. A very deep draw, a very large mold, a lot of sand, and it still hung up in the flask just like the smaller ones. We knew once we saw the molds that they make [that] making a bathtub mold may not be too difficult.

Barnes came to a similar conclusion when he saw the tank part being made:

> When I saw it, I didn't have my watch on to look at cycle time because that was the least of my concerns for this particular visit. It was [whether] you could physically make the unit. When they made a tank part that was

bigger than my desk and a deeper draw than an apron on the tub, I said, this thing will work. In my own mind it would work.

During this same period, APF's financial people were at the Hancock plant preparing cost comparisons between the current process and Process X. Also, a number of meetings were held with people from APF headquarters and Hancock to discuss Process X.

After the last of the U.S. visits was completed in January 1983, Barnes and Davis made a trip to Japan, accompanied by Hancock's plant manager, engineering manager, and tool and design manager. They saw the production lines at the Japanese bathtub producer, as well as its automated production line for brass castings.

As soon as he arrived at the Japanese firm, Barnes was asked by people who were licensing Process X to sign a contract to utilize their expertise with the process. He declined, however, believing that APF could do it themselves. During the visit, the APF personnel received answers to approximately seventy-five questions that they had prepared in advance concerning such issues as tooling requirements and metal composition. The trip convinced the Hancock people, as well as Barnes, of the desirability of Process X. In fact, their estimates of the potential rate of return on investment in Process X were doubled, based on expected labor reductions, material savings, and a lower percentage of product loss.

Barnes soon obtained quotes from Samson for Process X, which included an agreement to give APF a two-year contract, exclusive in North America. That is, if the agreement were signed, Samson would not license the process to anyone else to make bathtubs in North America for two years.

GAINING CORPORATE APPROVAL

Working with the division controller, Barnes completed his first proposal within a couple of months. The proposal contained background information and economic justification in terms of labor, material, and quality of the product. The next step was to secure approval from Dale Johnson, the president of APF, to proceed with the Process X justification. Barnes attributed the relative ease of gaining this approval to two things. First, Barnes had been hired by Johnson and had worked with him for several years; Barnes felt that he had credibility with Johnson. Second, he had kept Johnson apprised of developments throughout the investigation. Gaining Johnson's approval was a matter of a little conversation and some additional analysis.

Dale Johnson forwarded the proposal to his boss, Wayne Balston, a group vice president at Diversified, in May 1983. The proposal was returned for more work. Unbeknownst to Barnes, Diversified had changed its requirements for submitting capital projects; Barnes's Bennett-Watson format had become obsolete. Additional analyses, such as internal rate of return (IRR) now had to be run and added to the proposal. Once this was accomplished, the proposal could be resubmitted.

In the summer of 1983, Balston made a cursory presentation of the Process X proposal to his boss, Dennis Gallagher, the president and CEO of Diversified's manufacturing subsidiary. However, Balston recommended that any action on it be postponed. Thus the proposal remained dormant for several months, during which time Barnes repeatedly approached Dale Johnson to find out what had happened. Neither of them realized that Balston was about to retire and was therefore reluctant to push for anything as expensive and risky as Process X. Barnes and Johnson both believed that Balston was actively presenting their case to Gallagher during this time.

In October, Barnes presented a profit plan for APF to Dennis Gallagher. During the presentation, he mentioned Process X, the exclusive clause, and the importance of the process to APF. Because Barnes assumed that Gallagher was familiar with the proposal, his approach was matter-of-fact. To everyone's surprise, Gallagher acted as if he was only slightly familiar with the proposal. However, he also realized that APF was serious about it, noting that Barnes was backed by both Johnson and APF's marketing manager. He told them that "they would have to be patient"—that he would have to think about it some more. Barnes and the others for the first time learned of Balston's imminent retirement and the insincerity of his attempts to push for Process X.

However, Gallagher began to schedule some investigatory trips. He visited the Hancock plant and Samson, as well as the aluminum casting operation that Barnes and Davis had visited. Gallagher also had APF perform a great deal of additional risk analysis. Among other things, he had the potential effect of Process X on APF's china business analyzed, in addition to the cast iron analysis. Some of these evaluations had already been done, but they were reworked in greater depth. Barnes attributes the emphasis on risk analysis to the financial background of Diversified's corporate management.

In the meantime, Barnes and Davis met with Samson and decided to pursue a pilot line for Process X to prove the feasibility of making apron-style bathtubs, the largest piece made by APF. If bathtubs could be made with the process, then anything could be made with it. Even though the project itself had not been approved, Gallagher approved the pilot line. The agreement and purchase order with Samson were signed in December 1983.

In March 1984, Barnes visited Gallagher and reported the results of the pilot line operation. On the basis of that information, Gallagher gave his approval for the project and agreed to forward the proposal to the corporate level.

Dennis Gallagher then sent some of his financial people to APF and Hancock to provide details on items that Barnes had mentioned to Gallagher and to prepare a presentation for Gallagher to take to Diversified's board of directors. Barnes envisioned Process X as an innovation that would ensure the future viability of the plant and attract customers. (Barnes was probably the only one who believed in Process X's marketing potential at this time.) This spirit essentially disappeared in the version of the proposal prepared for Gallagher by the financial people. The new proposal was one page instead of ten and had a stronger financial flavor.

Gallagher had elected to justify Process X to the board strictly on the basis of its substantial cost reduction. The major savings were due to reduction in direct labor, but there also were savings in materials, lower cleaning room costs, and fewer rejections from enameling. Barnes believed that the project would not have been approved without presenting its financial benefits. In May 1984, Dennis Gallagher and Clifford Yates, chairman of Diversified, made the Process X presentation to the board of directors. The cost of the project was in excess of $5 million—the biggest capital investment ever addressed by the board. It was approved immediately. APF expected Process X to go on line sometime in 1985.

PRELIMINARY ANALYSIS

The APF case reinforces a number of themes that were illustrated in the other cases. Among these is the importance of being able to visualize the technology. In this case, visualization was achieved through plant trips. Both Elmer Davis and Fred Barnes placed great emphasis on their trip to the tank parts plant as a crucial step toward their determination that Process X could be used for bathtubs.

The importance of credibility is also underscored by the APF case. Because Barnes had good credibility with his boss, his recommendation to adopt Process X was forwarded with minimal delay. It is important to note that Barnes's credibility was not based on a previous success in implementing new technology, but rather on successes in labor negotiations and other areas. Apparently credibility is, at least to some degree, generalizable, reflecting a belief in a person's overall level of competence and trustworthiness rather than in his or her skills in a specific area.

This case also demonstrates the necessity of translating the anticipated technical benefits of a technology into financial or strategic benefits that can be appreciated by senior managers. Given that APF's parent corporation was not only unfamiliar with its specific business, but primarily involved with financial rather than manufacturing businesses, translation was especially important for this case.

The APF case also illustrates the importance of environmental scanning in uncovering innovative possibilities. Process X was brought to APF's attention because Elmer Davis attended a trade show. During two other trips, he made discoveries that progressively convinced APF that the process might be a feasible one for the Hancock plant. These somewhat unpredictable outcomes underscore the importance of such activities, which are often difficult to justify in advance.

Finally, the APF case reinforces the necessity of building a strong coalition across organizational boundaries, which I will call "solidarity." APF's initial overtures about Process X were scuttled by a manager who was close to retirement. The way in which the message was finally delivered to the executive in charge of Diversified Corporation's manufacturing subsidiary was as important as the message itself. As Fred Barnes put it, "When he [Gallagher] saw we were strong and unified...in an executive group...he started the ball rolling."

POSTSCRIPT

After APF had gained approval for Process X, one of their competitors called Samson to explore using the process for producing bathtubs. Of course, the exclusivity clause in their contract with APF prevented Samson from doing business with APF's competitors for some time. A number of this competitor's former customers contacted APF after hearing about Process X. Even the competitor has approached APF about making tubs for them, although APF has not followed up on this proposal to date.

APF has experienced substantial difficulty in implementing Process X at the Hancock plant. As of early 1987, it is trying to complete the debugging process in order to move into full production.

8 MONUMENTAL BUILDING SUPPLY
The MRP II Decision Process

Monumental Building Supply (MBS) is a wholly owned subsidiary of The Reston Corporation. From its beginnings in a garage in the 1950s, it has grown to be a major player in the market for aluminum windows and doors as well as aluminum extruded shapes for construction. MBS has three facilities: its headquarters and main plant in Shaler, New Jersey; a plant in Illinois; and a recently completed plant in Arizona. The company was purchased by Reston, a textile and chemical manufacturer, in 1970. (A partial organzation chart for MBS is included in Figure 8-1.)

Reston's annual sales are in the $1.5–2 billion range. MBS, the Nelson Company (an aluminum company whose total business is extrusions), and another smaller window company form Reston's Building Products Group. The Building Products Group comprises 10 to 15 percent of Reston's sales volume; MBS contributes slightly less than half of the group's sales. This is the story of MBS's attempts to gain corporate approval for a Manufacturing Resource Planning (MRP II) system. It is based on thirty interviews with people who were involved in the decision process (see Table 8-1) and discusses events occurring over the period from 1974 to 1985.

THE FIRST SYSTEMS (1974–1980)

In 1974, Andrew Gordon, a Reston veteran, was sent to MBS as vice president of operations. Gordon's background was in industrial engi-

Figure 8–1.   Partial Organizational Chart—Monumental
Building Supply.

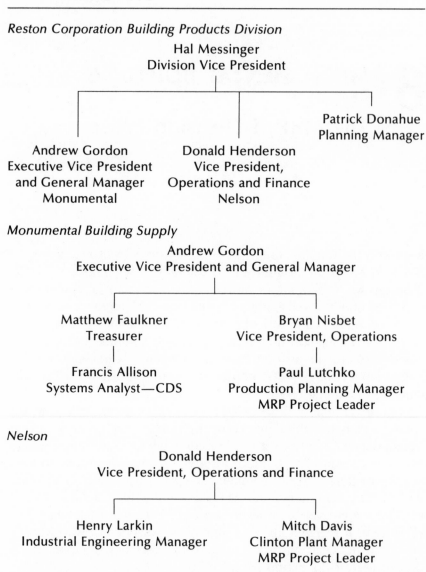

Reston Corporation Building Products Division

Hal Messinger
Division Vice President

Andrew Gordon
Executive Vice President
and General Manager
Monumental

Donald Henderson
Vice President,
Operations and Finance
Nelson

Patrick Donahue
Planning Manager

Monumental Building Supply

Andrew Gordon
Executive Vice President and General Manager

Matthew Faulkner
Treasurer

Francis Allison
Systems Analyst—CDS

Bryan Nisbet
Vice President, Operations

Paul Lutchko
Production Planning Manager
MRP Project Leader

Nelson

Donald Henderson
Vice President, Operations and Finance

Henry Larkin
Industrial Engineering Manager

Mitch Davis
Clinton Plant Manager
MRP Project Leader

neering. He had previously worked with a paper company that was also
a Reston subsidiary. Gordon saw his role at MBS as "bringing some
modern organizational thinking and management to Monumental, a
Reston perspective."

Table 8–1. Important Participants—Monumental Building Supply.

Allison, Francis   CDS Representative to Monumental, on MRP team

Benetton, Bruce   First Production Planning Manager at MBS

Burkhart, George   President, Reston Corporation

Davis, Mitch   MRP Project Leader, Nelson

Dixon, Warren   CDS Representative to Nelson

Donahue, Patrick   Planning Manager, Building Products Group, Reston

Edwards, Blake   CDS Executive, on Monumental's MRP Steering Committee

Faulkner, Matthew   Treasurer, MBS

Gordon, Andrew   Executive Vice President and General Manager, MBS

Henderson, Donald   Vice President of Operations and Finance, Nelson

Jankowski, Peter   Member of Monumental's MRP Team

Jarrell, Christopher   Head of Central Data Systems (CDS)

Jefferson, Blair   System Analyst at MBS

Jennings, Bob   Second Production Planning Manager at MBS

Klebe, Butch   Member of MBS's MRP Project Team

Larkin, Henry   Industrial Engineering Manager, Nelson

Lutchko, Paul   Third Production Planning Manager and MRP Team Head, MBS

Messinger, Hal   Vice President for Building Products Division, Reston

Nisbet, Bryan   Vice President of Operations, MBS

Washko, Steve   Member of MRP Team, MBS

Williams, Greg   Treasurer at Nelson, former MBS MRP Team Member

---

Soon after he arrived, Gordon established a centralized production planning operation. Production planning had previously been the responsibility of the plant managers. Bruce Benetton was promoted from head of industrial engineering and became the first head of production planning in 1975. He was charged with developing — at a minimum — a classical paper-based production planning system. Bob Jennings was brought in from MBS's Illinois plant to be assistant manager in the area and was given responsibility for day-to-day operations.

Benetton and Jennings both had interest in computer systems; they were not at all interested in developing a paper-based system. Blair Jefferson, an MBS systems analyst, agreed. The three of them knew that they could not wait until Reston's Central Data Systems (CDS) group developed a computer system, a process that would have taken several years. Furthermore, they were not impressed with the capabilities of the vendor software that was available at the time. Thus they set out to design the most crucial systems themselves.

One of the first systems they developed was a standard cost system. Bruce Benetton described the effort necessary to devise this and other early systems:

> We developed and tested a series of little programs, doing experimentation, working until midnight night after night. My wife wanted to divorce me; she was convinced I was seeing someone else. We wrote programs and used the mainframe illegally to test the system. We had friends who worked the night shift in Jacksonville [the computer center], who helped us get things done. CDS used to occasionally destroy our data bases, and we'd have to recreate them. They wouldn't know what it was, so they'd get rid of it.

Most of the work on the standard cost system had been completed by the time Benetton was promoted in 1975. At the end of 1976, the project was turned over to CDS to complete.

Another of the group's projects was the warehouse, a staging area where finished products awaited shipping. There was no system for locating any material in the warehouse; thus products were constantly being misplaced, and shipping dates were often missed because so much time was involved in looking for orders. One of the ways in which MBS tracks its performance is by determining the percentage of orders that are shipped on time and complete. In the mid- to late 1970s, this figure had slipped to under 50 percent. As Matthew Faulkner, MBS's treasurer, described it: "We ran what we thought was a very efficient, very effective manufacturing operation. We felt that no one could really hold a candle to us in terms of the quality of the product we manufactured. We just had a hard time getting it out the door."

In 1978, MBS constructed an 80,000-square-foot warehouse and instituted a manual location designator system to keep track of its products. The system was automated the following year with a Data

General minicomputer. This system—the first on-line computer system at MBS—improved performance dramatically. The number of orders shipped on time and complete recovered to about 90 percent within a year.

By 1979, however, it became apparent to Benetton and Jefferson that their attempt to control manufacturing operations with a set of pieced-together programs was not a good long-term solution. Coordinating the programs with one another and exchanging data with the corporate data processing center in Jacksonville had become an enormous task. After deciding that some form of integrated system would be necessary, they persuaded Gordon (who was now executive vice president and general manager) and Faulkner to include an MRP system in the data processing plan.

In the meantime, however, there were pressing order entry problems to be dealt with, and an additional stand-alone order processing system was developed to solve some of them. The project was initiated at MBS, but most of the work was done by CDS personnel. Benetton had originally suggested using vendor software for the order processing system, however, he was unable to convince CDS that this could be done. Work on the system was begun in 1980 and completed in 1983.

CHALLENGES TO MANUFACTURING (1980–1983)

Personnel changes, a failed presentation, and business conditions delayed MBS's investigation of an MRP system during this period. Benetton left MBS in 1981; Bob Jennings was promoted to production planning manager. In an attempt to get the MRP effort moving again Jennings hired Will Gearhart, a salesman from Sperry Corporation, as the MRP project leader. By all accounts, Gearhart was more of a salesman than a technical person; he did not seem to fit in well at MBS. However, he was able to lead MBS in an evaluation of MRP packages, including Xerox, American Software Incorporated, and Management Science America.

Excitement about the possibility of finally pursuing an integrated system was building at MBS. In late 1981, a brief review of MBS's MRP efforts was scheduled as part of a meeting with Hal Messinger, vice president of Reston's Building Products Division, and some of Reston's corporate representatives. The presentation was slated for ten minutes, but Gearhart took over an hour, using fluorescent pencils and other flashy presentation aids. The presentation was a disaster:

its substance was overwhelmed by its style and length, and the participants resented being lectured to. It served only to destroy the progress MBS had made to date and substantially reduce its chances of getting corporate approval for further investigation.

In mid-1982, Bob Jennings was promoted to plant manager of MBS's Illinois facility. Paul Lutchko (formerly the industrial engineering manager) replaced him in production planning. Lutchko and Gearhart did not see eye-to-eye on MRP. Gearhart left MBS within a few months. Before Lutchko could attend to MRP, he needed to familiarize himself with the production planning operation and provide some leadership to the area. Thus it was almost a year before MRP received any new impetus.

In the meantime, a number of factors had conspired to make MBS's manufacturing task much more difficult. First, MBS had entered the commercial and "rehab" (replacement window) markets. Second, the company had decided to extend its market from its traditional bailiwick in the mid-Atlantic and Northeast to include the Sunbelt and Florida. This move was based on the realization that half of all new housing starts were occurring in three states: Florida, Texas, and California. Third, high energy costs had created a new market in energy-efficient windows, which MBS was trying to enter. While all of these factors represented opportunities for MBS, they resulted in a greatly increased number of product offerings. In the early 1980s, the number of basic product lines offered by the firm increased from approximately fifteen to forty, with greater variation within each product line. Such product proliferation subtantially increased the complexity of manufacturing control at MBS.

At the same time, the homebuilders that represented MBS's main customers were reeling from increasingly higher interest rates as well as the burdensome inventory costs they represented. These customers responded as many firms did at this time—they demanded more frequent smaller deliveries and shorter lead times between delivery dates and the time the products were actually needed. This in effect forced MBS to carry inventory that was formerly carried by its customers. In addition, reliable on-time and complete delivery was being placed at a premium; business was going to companies that could accomplish this. MBS was being forced to compete with regional producers that only had a few products and could provide reliable delivery within two weeks, as compared with MBS's four- to five-week cycle. Therefore, MBS was attempting to produce a much higher volume and wider variety of windows, in smaller lot sizes with shorter delivery schedules in order to provide the reliable delivery its customers were demanding. MBS's home grown information systems, which had performed adequately in the simpler world of the late 1970s, became rapidly over-burdened by new competitive demands.

Problems with inventory control, bills of materials, and capacity planning became increasingly severe. On-time and complete deliveries, which had risen above 90 percent with the introduction of the ware-house system, fell to between 60 and 70 percent at a time when the firm could least afford it. Products were constantly being expedited through the factory. Pressure from both the customers and the factory floor was building. Production planning would need to do something about the problems. The need for an integrated MRP system was becoming increasingly obvious to MBS's management.

THE AUGUST 1983 MEETING: A SURPRISE

In October 1982, at the urging of Lutchko, Peter Jankowski was hired to fill the assistant production planning post previously held by Gearhart. Jankowski's primary task was MRP implementation. Tip Johnson was brought in to handle the day-to-day production planning responsibilities. Lutchko was able to convince management that it was time to revive MRP. (Lutchko reported to Bryan Nisbet, MBS's vice president of operations.) In addition, Andrew Gordon and Matthew Faulkner participated in the planning for the MRP effort. (As treasurer, Faulkner had functional responsibility for data processing.)

Lutchko and Jankowski began to make an assessment of just what needed to be done in the MRP area by conducting another round of vendor evaluations. Among the firms they looked at were ASI, MSA, and Cullinet. They found that MRP packages had evolved quite a bit in the five years since Benetton and Jefferson conducted their investigation. MBS became more convinced that one of these new packages would be the solution to many of their production problems. Jankowski and Lutchko attended a number of seminars on such topics as software evaluation, system implementation, and so on.

One of the seminars was offered by a man named Terry Schultz of The Forum Limited. Lutchko had been impressed with this organization, and in June 1983 invited Schultz to give a "kickoff" presentation for MBS's renewed MRP effort. Schultz was on-site at Shaler for two days and his seminar was attended by twenty-five people from MBS, Nelson, and CDS. For Lutchko, Schultz only reinforced what he had already heard about MRP implementation, but he wanted this information to be more widely shared in the organization.

At about this same time, an MRP project team was formed that included both Lutchko and Jankowski. In addition, Francis Allison, a systems analyst from CDS in Jacksonville, joined the team. Allison had been involved in the development of the order processing system that CDS had done for MBS. He had been visiting Shaler on a regular basis for several years. Finally, Greg Williams, MBS's comptroller, provided the accounting perspective. (Williams's tenure on the committee was brief, however, as he accepted the treasurer position at Nelson in mid-summer 1983.)

MBS held a meeting in August 1983 to attempt to clear up some problems with the order processing system. In addition, it planned to deliver a progress report on MRP to two key individuals: Hal Messinger and Christopher Jarrel. As noted, Messinger is vice president of Reston's Building Products Division. He was the first person whose approval would be necessary for any capital project — including MRP. If Messinger did not support the project, it would go no further. MBS's management, however, was reasonably confident that Messinger was in support of its MRP plans.

Christopher Jarrel was the director of CDS. He would have to approve any computer systems implemented by MBS. Reston, like most companies, traditionally operated in a centralized data processing environment; Jarrel was known for his resistance to decentralization of the computing function. In order for any MRP system to be effective at MBS, the project team felt a mainframe would be needed in Shaler. In addition, MBS was leaning heavily toward purchasing MRP software, as opposed to having it developed by CDS. This would also be an unusual move for Reston. While industry in general was moving in the direction of distributed data processing and vendor software, MBS's management was not at all sure that they would be able to overcome Jarrel's traditional resistance in these areas.

Preliminary economics for the MRP project were presented at the August meeting. MBS anticipated benefits in several areas: reduced inventories, greater manufacturing efficiency, reduced purchasing costs, and greater sales and profits. Costs, including hardware, were estimated to be in the range of $1.6 million over five years. Payback was estimated to be within four years. The return on capital was expected to be around 19 percent.

MBS planned to use Messinger's support for MRP to help sell Jarrel at the meeting. Instead, they were surprised to find Jarrel in support of the project. Messinger, however, was unconvinced. There were at least two reasons behind Messinger's hesitancy. First, he was not sure that the business could support such a large investment; MBS was projecting $600,000–700,000 for a mainframe, in addition to approximately $1 million for software. The recessions of the early 1980s had dramatically driven down the number of housing starts, which had, in turn, depressed the window market. Because MBS is a volume-sensitive operation, profits were way down. Messinger probably also suspected that the actual dollar figure required to implement MRP would be quite a bit larger than what was being projected. Like many other firms, MBS had historically experienced substantial cost overruns on data processing projects.

Messinger's second major objection was that MBS's plans had no provision for participation by Nelson, its sister company in the Building Products Group. He pointed out that MBS and Nelson both had done customer surveys and both had concluded that on-time, complete delivery should be their prime manufacturing objective. Both were targeting faster throughput time and decreased inventory as important goals. Andrew Gordon recognized the similarity of goals between the two companies: "We had made a number of market surveys, both Nelson and ourselves, and we had come to the same conclusions, that we could be on the leading edge, we could beat our competition if we could improve our ability to ship earlier, and to ship on time and complete as promised." Despite these common goals, each company was proceeding independently with plans for computer systems that would help achieve these goals. For example, Nelson and MBS each developed its own order processing systems with CDS; had they cooperated in developing one system, the total cost would have been quite a bit lower. Similarly, each had purchased a different CAD system.

Messinger and others in his group also realized that substantial efficiencies could be achieved by having MBS ship some of Nelson's extrusion products, and vice versa. As it stood, MBS was trucking products down the East Coast from New Jersey to Florida; Nelson was sending them up the East Coast from Alabama to the Northeast. In order to work together, however, the two companies would need some common systems. Messinger was concerned that MBS's plan to implement an MRP system would drive the two companies further apart. He indicated that if MRP were to proceed at all, it would have to be a "common thrust" between MBS and Nelson.

THE DECEMBER 1983 MEETING: ANOTHER SURPRISE

MBS's MRP leaders (Gordon, Faulkner, Nisbet, and Lutchko) were again disheartened by another setback. In the fall of 1983 Lutchko was given the responsibility of re-presenting the case to Messinger. He drew up a number of reports, which were reviewed by Gordon, Nisbet, and Faulkner, but no one was sure what arguments would be needed to convince Messinger.

Finally, they decided to re-contract Terry Schultz to help sell MBS's approach. Given Messinger's insistence that MRP would have to be a joint effort, it would also be necessary for MBS to convince Nelson. This predicament would be compounded by the fact that Nelson's incremental approach to manufacturing systems was in conflict with MBS's total system proposal and MBS's desire to use an outside vendor conflicted with Nelson's desire to continue to use CDS to develop its systems. In addition, MBS and Nelson had a history of competition rather than cooperation; thus there might be some reluctance on the part of Nelson to follow any MBS-directed lead.

MBS feared that Messinger might try to unify the companies by forcing it to use some of Nelson's CDS-developed systems. Two systems in particular were of concern: PIS (production information system) and PMS (profit measurement system). MBS felt that these systems did not fit the needs of their assembled products as well as those of the one-piece extrusions produced by Nelson. (In fact, most of MBS's manufacturing control problems were occurring in the assembly area, not in extrusion.) In addition, MBS believed that it would be difficult to fit an integrated MRP system around these two systems. MBS needed to let Hal Messinger know that the Nelson systems would not be appropriate for MBS, without appearing to criticize Nelson for having implemented them.

In December 1983, a meeting was held at Nelson's headquarters in Clinton, Alabama. It was attended by the MBS and Nelson management teams, including Donald Henderson, Nelson's vice president of operations and finance. Christopher Jarrel and several others from CDS also were in attendance, as well as Messinger and Patrick Donahue, a planning manager in the Building Products Group. As expected, Schultz gave a convincing presentation of the benefits of integrated MRP systems, and MBS reiterated its assumptions and proposals in this direction. MBS was hoping to get approval to conduct a feasibility study for MRP.

To everyone's surprise, Messinger announced that he was in favor of MBS's total system approach and that he would authorize the investigation to continue through the vendor selection stage. He also reiterated his sentiment that it should be a joint project between Nelson and MBS. He then directed Nelson to "get on board early" by actively participating in the investigation. While MBS was pleased with Messinger's endorsement, it was concerned with the necessity of working the project through with Nelson. The only support MRP enjoyed at Nelson came from relatively low-power people: Greg Williams, who had come from MBS, and Warren Dixon, who had just arrived from CDS. MBS's concerns were somewhat alleviated the following month when Messinger clarified that he did not necessarily mean that the two companies had to embark on the study together; he merely required that there be some communication.

MBS TRIES TO GET STARTED (1984)

In early 1984, Francis Allison relocated to be at Shaler full time. This move was part of a CDS experiment to have its systems analysts located at various manufacturing sites. Allison was one of the first to volunteer for this effort. Having visited MBS on a monthly basis for three years in connection with his role as designer of the order processing system, Allison was familiar with its operations. By this time, Lutchko had been able to secure management's support for nearly full-time commitment to the MRP project for himself, Peter Jankowski, and Allison, and half-time commitment for Butch Klebe (Williams's replacement).

Lutchko was very eager to proceed with the project. It had been almost two years since he had taken over production planning, and MRP still appeared to be a long way off:

> I'm getting tired of making excuses, of hearing manufacturing people saying "You've got to help us out." We've held the MRP carrot out there to those people for so long, and they keep biting. Sooner or later they are going to bite our hands; they're not going to believe it. My credibility is going down the tubes. . . . I used to say "MRP can solve [your problems]." Now I've quit saying it because they say, "MRP? When? Before I retire?" Now I can't guarantee MRP. It was tough to tell them to wait, but it's even tougher to say nothing.

In January and February 1984, as Allison's work on the order processing system was drawing to a close, Lutchko and Jankowski undertook a project to regain some discipline in the use of the warehouse system. Believing that people had become sloppy in their use of the system, they held a series of meetings intended to restore the system's integrity. Inventory accuracy in the warehouse had fallen to the 60–65 percent range. Since discipline would also be necessary in the implementation and use of an MRP system, Lutchko and Jankowski saw this project as something of a test run. Their efforts were rewarded with an improvement in inventory accuracy to almost 90 percent in the first month of reform and 95 percent a few months later. More important, they believed that they had finally built momentum for pursuing MRP.

Consultants also began to play a role in the justification process. A Coopers and Lybrand (Reston's auditor) representative had been suggesting to Andrew Gordon that someone from his company's consulting branch could help MBS in its preparation for MRP. Meanwhile, Randy Randazzo, an MRP consultant from Arthur Anderson, was visiting Henry Larkin, Patrick Donahue, and Warren Dixon (Allison's equivalent at Nelson). Donahue suggested to Gordon that Randazzo also visit MBS; Gordon agreed on the condition that someone from Coopers and Lybrand would also be interviewed for a possible consulting role. The visits took place during March and April 1984. MBS was extremely impressed with Randazzo, whom they saw first. (It was not at all impressed with Coopers and Lybrand's consultant.) Randazzo became MBS's obvious choice for an MRP consultant.

A WATERSHED FOR MRP

MBS's momentum was soon endangered by an issue that temporarily polarized their management team and called into question the future path of the MRP effort. The issue was whether to implement a software package called "Purchasing and Materials Management" (P&MM).

Several years back, Reston had decided to purchase an inventory management package for use in various corporate facilities. MBS was asked if they wanted to participate in the search. It agreed on the condition that the search be limited to vendors offering complete MRP packages that included inventory management. CDS agreed to MBS's stipulation. MBS was represented in this process by Will Gearhart and a CDS analyst named Roger Reed. Reston arrived at a decision to purchase ASI's P&MM system. (It was noted that a much less expensive system could have been bought absent the constraint imposed by MBS.) The system was implemented at a chemical plant in Peachdale, Georgia, as well as in Nelson's maintenance department. Huge benefits were being claimed, particularly at Peachdale.

In spite of having participated in the study, MBS initially passed on the opportunity to implement P&MM, contending that they were close to beginning MRP, which would encompass inventory management. In the fall of 1983, however, MBS committed itself to implementing P&MM in the first quarter of 1984; substantial pressure was coming from both Hal Messinger and CDS to make good on that promise. Messinger had been behind the implementation of P&MM at Nelson. The immediate impetus forcing Monumental to decide whether to implement P&MM was influenced by the long-range data processing plan, due in February.

Matthew Faulkner felt that MBS should follow through on its promise to implement P&MM for a number of reasons. First, and probably most important, he believed that MBS lacked the credibility in implementing packaged systems that would be necessary to gain acceptance for MRP. Implementing a more limited package like P&MM would help MBS to acquire this credibility. Second, P&MM would allow MBS to discover whether CDS's mainframes in Jacksonville would be able to provide the response time and uptime to support this type of on-line system (almost everyone at MBS thought they would not). Such evidence would aid MBS in its request that a mainframe be placed at Shaler for the eventual MRP system.

There was also some feeling that ASI was gaining a lot of momentum at Reston. MBS would therefore be placed in a defensive position if it recommended anyone else for MRP. Faulkner claimed that P&MM would give MBS the expertise to either accept or reject ASI, based on extensive experience with one of its central MRP modules. Finally, P&MM would simply allow MBS to gain some experience in project management and system implementation on a smaller scale than MRP. Faulkner described this as "testing in shallow water before you jump into the deep with a swift current." He expressed his sentiment as follows:

I just don't see the corporation telling us go ahead and spend that money on something else when they've got a commitment to ASI already and the people using it are very pleased with it. It's been difficult for us right along, and I think that it's going to be more difficult for us in the future to turn our backs on P&MM, because the corporation is getting more and more deeply ingrained in that system.

Faulkner's view was not shared by Paul Lutchko, who felt that implementing P&MM would be devastating to MRP. Latchko marshalled a number of arguments to support his contention that P&MM should not be implemented. First, he believed that MBS already had adequate experience in implementing systems, which included the warehouse and order processing systems. Second, he said it was obvious that CDS's remote mainframes could not provide the kind of support needed for MRP; MBS did not need to spend $70,000 on P&MM to find that out. Third, he agreed with Faulkner that the pressure to use ASI for MRP existed, but stated that implementing P&MM would only increase that pressure.

Lutchko did not want to be "locked into ASI," explaining that its MRP system was "good in some areas and bad in others." And he certainly did not want to fit another vendor's MRP system around ASI's P&MM, a combination he considered "a data processing nightmare." He also questioned the claims coming from Peachdale, which was primarily an R&D facility, and their relevance for an operating unit like MBS. In summary, Lutchko felt that P&MM "would not address the real problems in manufacturing" and that eventually replacing it with an MRP system would "tear out the heart of the people on the floor."

The issue came to a head in March 1984 during a series of meetings involving Gordon, Faulkner, Lutchko, and Allison. While all of the participants agreed on the ultimate goal of arriving at MRP, they disagreed on the approach. Gordon was particularly anxious to get started on MRP but uncertain as to whether P&MM would hasten or prolong the process. The meetings focused on whether MBS would be able to sell MRP to the corporation in general and Hal Messinger in particular without first implementing P&MM, and whether MRP vendor selection could be accomplished before the P&MM decision was completed. Gordon had been told unofficially by Randazzo that MRP vendor selection could be done in less than four months with Arthur Anderson's help. If this were true, then MBS could either choose ASI for its MRP vendor (and implement P&MM) or have stronger arguments against ASI and in favor of another vendor.

Gordon decided to delay the final decision until he had received a formal report from Arthur Anderson and had a better sense of the political implications of introducing P&MM. The long-range data processing plan was submitted to CDS by Francis Allison. It contained two alternatives: one that involved implementing P&MM and then MRP, and one in which MBS would proceed directly to MRP. Gordon said that he would make his decision after the long-range planning meeting was held.

The final resolution of the P&MM issue was delayed for two reasons. First, Donald Henderson had promoted Henry Larkin to industrial engineering manager at Nelson, charging him with continuing responsibility for studying the on time and complete problem. Henderson asked Gordon to delay his decision until Larkin had completed his own study. Second, Christopher Jarrel had resigned as head of CDS, and no successor had yet been named. (It was rumored that Jarrel's resistance to decentralization of data processing had led to a parting of the ways between him and Reston's Executive Committee.) CDS would be at a standstill until a successor was named and its future direction became clear.

In the meantime, Faulkner, Nisbet, and Lutchko took the opportunity to visit the two P&MM sites, where they could assess for themselves the benefits that were being claimed. During this trip, Nisbet, who had been away during the earlier discussions about P&MM, came out strongly against installing P&MM for purely political reasons. Nisbet, Faulkner, and Lutchko agreed that they should recommend that MBS "do the right thing," that is, implement MRP, and allow Messinger to tell them otherwise if he wished.

Faulkner, Nisbet, and Lutchko also held discussions with Nelson during this trip. Lutchko came away concerned with what he perceived to be a lack of commitment to MRP within Nelson — despite Messinger's pronouncement at the December meeting. He felt it would be helpful for people on the (Jacksonville) Building Products Group staff to be more familiar with MBS's MRP concepts. He asked Patrick Donahue, planning manager for the group, to attend the "Business Requirements Planning" seminar offered by The Forum Limited. Donahue agreed and attended along with Henry Larkin. Donald Henderson and Warren Dixon followed a month or so later. As a result of attending the seminar, both Donahue and Larkin appeared to grow in their appreciation and support for MRP.

At the end of May, Andrew Gordon fulfilled Donahue's request to spell out MBS's plans by sending him a document entitled the "Monumental Products MRP Project Plan." This document was very detailed and included sections on scope, objectives, costs and benefits, organization, and background information. This was the first time that MBS had displayed the degree of thought and planning that had gone into their MRP efforts.

On another front, Gordon received Randazzo's recommendations concerning MBS's approach to MRP. Randazzo had been asked to comment specifically on the wisdom of MBS's current official proposal: to implement P&MM while simultaneously investigating MRP. He came out very strongly against this plan, having concluded that the MRP vendor selection would take only about four months, at which time ASI would be either confirmed or rejected as the vendor. A copy of Randazzo's letter was sent to Patrick Donahue in early June, and Gordon pressed Donahue to share the conclusions contained therein with Messinger.

Many of these threads came together at a June 1984 meeting in Shaler. In addition to MBS management personnel, Nelson representatives and Donahue attended. Both Nelson and MBS presented their progress and plans concerning systems to resolve their problems with on-time and complete delivery rates. MBS felt that it needed to convince Donahue — and, through him, Hal Messinger — of its need to proceed directly to MRP. Donahue, however, surprised the participants by delivering Messinger's decision that MBS should in fact go ahead with the requirements definition and vendor selection for MRP.

Messinger agreed on the conditions that MBS not neglect other projects, such as completing the new plant in Arizona, and that it select a system that could also be used by Nelson. Donahue was given responsibility for coordinating the efforts of the two companies, and Lutchko agreed to communicate with Donahue once a week.

GETTING STARTED AND CHOOSING A CONSULTANT (MID-1984)

MBS's MRP effort finally began in earnest. The project team initiated meetings in June. Steve Washko, the expert on MBS's standards and quoting system, was released to spend 30 percent of his time on MRP.

Meetings were held about twice a week throughout June and July. The team reviewed books and videotapes about MRP, prepared to interview MRP users, and began to work on a "conceptual design report," a high-level description of the proposed system. The report was prepared to educate both MBS and Reston, to provide background information a consultant would need, and to provide documentation to satisfy CDS that a proper feasibility study had been done. The team's main problem during this period was finding sufficient time to work without neglecting their other responsibilities.

On July 17, the first meeting of the MBS and Nelson "teams" was held. Henry Larkin was the only person at Nelson who was actively working on MRP, although Donald Henderson and Mitch Davis acted as advisers. Both companies reviewed their progress. Arthur Anderson and Coopers and Lybrand were identified as prime candidates for a consulting role in the project. The Oliver Wight Company, an MRP education firm, also was asked to submit a proposal for the job but had not yet replied.

During the meeting, Patrick Donahue reemphasized the importance of the MRP investigation to both firms. The participants felt that Donahue's statement was another in a series of signals sent by Messinger that he had become committed to MRP. Messinger had informally told a number of individuals over the past few months that it was very important that MRP be implemented quickly, correctly, and inexpensively. Donahue noted in a memo summarizing the meeting that MBS was clearly moving faster than Nelson, and that Nelson should be concerned about this, given the fact that they needed to be in a position to make an informed choice when hiring a consultant.

The consultants participated in a meeting held in Clinton on July 31. Arthur Anderson, Coopers and Lybrand, and Oliver Wight all made presentations. Arthur Anderson's recommendation that it be involved part time, in an advisory capacity, was consistent with MBS's expectations. Wight recommended its standard approach — heavy on education and light on hands-on consulting. As noted, everyone at MBS was very impressed with Randazzo, Arthur Anderson's representative. He had already been instrumental in convincing Donahue and Messinger that MBS should forgo P&MM in favor of a broader investigation. Arthur Anderson, however, tries to sell its own (IBM) MRP package, so MBS was concerned that it might not be objective in evaluating other packages. Arthur Anderson also had no history of involvement with Reston and might therefore have limited credibility.

Matthew Faulkner had called Coopers and Lybrand prior to the meeting and told them as tactfully as possible that the individual initially sent to MBS might not be their best representative for this project. Coopers and Lybrand responded by sending the partner in charge of the Reston account along with its MRP expert, the national director of materials management to the meeting. By this time, MBS was leaning toward Coopers and Lybrand since its long-term relationship with Reston would provide credibility when the time came to sell the project in Columbus. Coopers and Lybrand would also be able to provide a "features and functions checklist," a detailed comparison of several MRP systems that would enable MBS to identify the best systems for its needs.

At the July 31 meeting, however, Coopers and Lybrand expressed its desire to send one full-time person and an additional part-time person to both MBS and Nelson. This was far over the level of involvement desired by MBS. After Coopers and Lybrand was challenged by MBS to justify the need for this level of participation, it came back with a proposal that corresponded more closely to MBS's wishes. Additionally, it agreed to let MBS choose the associate who would handle the job. Once MBS had identified a capable individual, they committed to using Coopers and Lybrand for the consulting role. Since Nelson was not yet committed to MRP, they were not convinced that they needed a true consultant; they planned instead to continue to use Oliver Wight as an educator.

MBS's project team continued to develop internal plans. Andrew Gordon distributed a "mandate memo" in late July, authorizing the project team to begin conducting their user interviews. These interviews continued through October. A draft of the conceptual design report was completed, and the steering committee met in early August to discuss it.

A significant business issue concerning the target lead-time for delivery had implications for MRP. Lutchko felt that if the target were twenty-one days, MBS could continue with the current batch-oriented physical system. But if the target were to be only fourteen days, a more customer-oriented system would be needed. MBS's management was reluctant to endorse one lead-time for all products and customers. Unfortunately, since it did not have a marketing department per se, no one knew which customers or product markets required shorter lead times. Hank Baskins, who had recently been hired to fill the void in marketing, agreed to take on the market study that would help to answer these questions.

NELSON'S POSITION AND PROGRESS

Nelson began to move toward MRP in early fall 1984. In his final report on the on-time and complete issue, Henry Larkin had favored an integrated system solution such as MRP. He arranged for Hal Messinger, Donald Henderson, John Henderson (vice president of extrusion sales at Nelson), and Patrick Donahue to attend an Oliver Wight seminar in mid-September. After the seminar, these four individuals decided that they should comprise the steering committee for Nelson's MRP effort. As such, they also established a project team, consisting of Larkin, Mitch Davis (the assistant plant manager), Tom Wholey (shipping superintendent), Warren Dixon (from CDS), and Bob Dowell (from the sales organization). All of these people were assigned to participate in the MRP investigation in addition to their other duties. Several of them were sent to the Oliver Wight course. The team began to meet at one- to two-week intervals.

Nelson was in a difficult position with respect to MRP. On the one hand, it could see the advantages that such a system could provide. On the other, it had invested significant capital in various in-house systems and had recently rewritten PIS (one of its major packages) for use on the Data General computer. This rewritten system was scheduled to come on-line in fall 1984 — MRP would threaten its survival. All of this investment was a sunk cost when viewed from a purely objective basis. However, it would be difficult for any firm to turn away from systems in which it had such substantial financial and personal investment.

MBS continued to share information and plans with Nelson. In October, Lutchko, Gordon, and Les Shindleman (the Coopers and Lybrand consultant) went to Clinton to visit with Messinger, as well as with Nelson's management and MRP team. MBS was expecting to begin looking at vendor software by the end of the year and wanted to know in advance whether the differences in pace between the two project teams would hinder their progress. Donald Henderson reiterated his earlier statement that MBS should take the lead on the project and indicated that Nelson would not hinder its efforts to proceed.

THE NOVEMBER 1984 MEETING AND ITS AFTERMATH

In November 1984, the user interviews were completed and several personnel changes were made. Peter Jankowski was promoted to industrial engineering manager and moved off the MRP project team. Butch Kleve was promoted to system analyst, thus decreasing his involvement in the project. Steve Washko was made a full-time team member.

By mid-November, the team had also completed the organizational impact report which included a proposed conceptual design for the MRP system. On November 20 MBS's steering committee and some additional people from Nelson met to discuss both the report and some management issues that needed to be addressed for MRP to work. These included the need for forecasting, some organizational changes, including the creation of a department with responsibility for inventory, and changes to existing software packages such as the warehouse and order processing systems.

While the meeting itself went well, the MRP team later received a shock in the form of a memo from Warren Dixon, the CDS analyst working on Nelson's team. While the memo contained some positive comments on MBS's report and presentation at the meeting, it was dominated by challenges and questions regarding MBS's approach and progress to date. The challenges centered around four issues: the lack of detail provided, lack of knowledge of the project team, the heavy use of the consultant, and the lack of coordination with Nelson. The memo was addressed to Roger Reed at CDS; copies were distributed to Donald Henderson, Mitch Davis, Paul Lutchko, and Blake Edwards (a senior manager at CDS who was on MBS's steering committee). The informal distribution of the memo was quite extensive.

MBS was very distressed by the memo because it represented a challenge to their credibility as MRP implementers, and the challenge was coming from an individual who had been regarded for some time as CDS's in-house MRP expert. Because Dixon had been involved in a disastrous MRP implementation at another company, he was extremely conservative about having all the details under control in any large-scale project. Lutchko sent a reply to Dixon, indicating that many of his concerns were based on a misunderstanding of MBS's intent. The meeting was not intended to provide the kinds of details Dixon was looking for; these would be forthcoming in a report to be released early the next year.

Dixon's memo was especially surprising in view of the fact that Edwards, who was a level up from Dixon within CDS, had publicly commended MBS's progress. People at Nelson told Lutchko privately that Dixon's views were his alone; they did not represent the general consensus. (Interestingly, Dixon later told people at MBS that some of his questions were really directed at Nelson; the memo was only his vehicle for raising these issues.) Although the memo created quite a stir, it ultimately did not endanger MBS's credibility or progress in implementing MRP.

PAVING THE WAY (EARLY 1985)

Throughout the end of 1984 and into 1985, MBS continued its work. Positive signals from Reston were received during the budget presentation — including the fact that MRP would be mentioned in the forthcoming annual report. Also, Reston's corporate director of productivity (who had attended the November meeting) had told MBS that he felt they were on the right track. The requirements summary report was completed in early February. The way was now clear for MBS's vendor investigation.

Things seemed to be moving ahead well. Paul Lutchko had released a memo in January that identified "milestone events" for the conclusion of the MRP investigation and established an aggressive schedule for its completion that called for the project to be submitted to Reston by mid-year. Nelson endorsed the schedule, again telling MBS that it would not stand in its way. Nelson seemed to have found its own MRP champion in Mitch Davis. Paul Lutchko was very conscious that the project's success would be influenced by whether Nelson was comfortable with the vendor selection process.

> Nelson is going to have to commit themselves to the selection that's made. Even though they're not in a position to issue their own [Requirements Summary], the whole purpose of us sitting down and doing all of this is for us to try to convince them that what we have here satisfies MBS but it probably satisfies a lot of their needs too. They shouldn't be real nervous about the vendor selection process. That if they agree with most of these things and if we can find a vendor who does most of these things,

then there may be a good chance that the vendor can do for them what the vendor can do for us.

MBS was provided with two opportunities in early 1985 to pave the way for the eventual project submission. In February, Nelson was preparing to invest at least $25,000 in training for MRP. Donald Henderson felt that he should not devote this amount of money to MRP unless he received some signal from Reston's Executive Committee, which would eventually have to approve the project, that it was viable. A meeting was held with representatives from MBS, Nelson, the Building Products Group, and the Executive Committee in Jacksonville. Paul Lutchko and Mitch Davis presented their progress to date and preliminary economics, and Donald Henderson and Andrew Gordon discussed the business rationale for the project. The meeting was a success; the committee encouraged the two companies to proceed with MRP.

Another opportunity was afforded in March, when George Burkhart, Reston's president, visited MBS in Shaler. The project team took the opportunity once again to review their progress and plans with a key member of the Executive Committee. Burkhart was very encouraging; he concluded with the statement, "It sounds great, let's get going."

INVESTIGATING VENDORS (MID-1985)

MBS and Nelson were by this time well into vendor investigation. The team had spent a day in January with Shindleman and another Coopers and Lybrand consultant in order to draw up an initial list of potential vendors to be narrowed down to six candidates. It was relatively easy to identify the twelve vendors that offered IBM mainframe software written in Cobol. Narrowing the list to six, however, proved more difficult.

One of the reasons MBS had hired Coopers and Lybrand was because of its "features and functions" checklist. However, MBS found that the list did not fit well with its own approach. Several features were often included within one line item, and other important functions were not included at all. In addition, business or "political" factors were not included — that is, factors that would be important to the managers who would have to approve the project, including the financial stability of the vendor, the number of installations the vendor had in place, and the company's reputation within Reston in general and CDS in particular.

The features and functions checklist arrived at an overall score for each package by assigning a weight to each feature needed, a score to each feature of each package, and multiplying the scores by the weights. Francis Allison in particular felt that this method would not lead to the kind of decision that was needed:

> You've got some "have-tos," and you've got some "really-wants," and you've got some "niceties," okay? But when you jumble them all together, you could have a package that has ten "has-tos" missing. If they have everything else, they might come out real good. . . . I like to go through the list as a person, and . . . have consensus among the group, and not have some . . . mathematical equation make my decision. I'm not saying the approach is bad, but I don't agree with the approach . . . based on some nontechnical factors which are not really in there. Political factors they're called. . . . Are they a solid company, are they an accepted company to deal with in terms of Reston, those kinds of things. . . . You don't want to do the mathematical equation and come up with something that you can't sell or that you're going to spend years selling. Let's get a good package that we can sell.

In spite of its reservations about the checklist, MBS felt that the consultants were indeed helpful in providing information about vendors. As a result of their meetings with the consultants, the team succeeded in dividing the original twelve vendors into two groups—the top six, and a second six that were less likely to be chosen. A hardware expert from CDS then was enlisted to contact each vendor and obtain answers to a list of technical questions concerning operating systems and data base structures. In addition to the fact that it truly needed this information, MBS hoped that CDS's continued involvement with the project would diminish the need to sell it to them when it came time for approval.

On the basis of their investigations, MBS and CDS narrowed the set of vendors to four: MSA, Comserve, Cullinet, and ASI. Each vendor was invited for a one-day meeting, which would include the vendor's presentation and a question-and-answer session. Representatives from Nelson attended all the presentations, some of which were held in Clinton.

The first presentation was made by Cullinet. The MBS team was very impressed with this package. In fact, the team was apprehensive that all of the packages would be as good, which would make their decision very difficult. Any such fears were alleviated the following day with MSA's presentation, however. MSA was immediately dropped from further consideration on the basis of a poor presentation and missing elements in its system.

The Comserve and ASI presentations were held in Clinton. Because it had many installations and the appropriate data base management system, Comserve had been the early favorite. The team thought that the system was good but felt that the sales personnel had exaggerated its capabilities. For example, they claimed that the system was completely on-line (an MBS requirement), but it was actually in the process of being converted from batch to on-line. ASI made a good presentation, and was seen as having a very sophisticated system as well as very good salesmen. ASI was handicapped, however, by the fact that the P&MM software package—which MBS had considered and which Nelson had installed—had now acquired a bad reputation. CDS in particular had been soured on ASI.

Two-day sessions were held in May with each of the remaining vendors—Cullinet, Comserve, and ASI. These sessions were held at the vendors' facilities and were again attended by both MBS's and Nelson's project teams. On the basis of these sessions, the field was narrowed to Comserve and Cullinet.

The last and most extensive stage of vendor evaluation thus focused on these two firms. Additional visits were made to the two firms, as well as to user firms that had been provided as references. A number of phone calls were made to other user firms. The two project teams held many joint sessions to discuss this information. MBS had been leaning in the direction of Cullinet, but was trying to be objective about the strengths and weaknesses of both vendors. After this last round of interviews, however, the project team reached consensus that Cullinet would be their choice.

MBS's management and MRP teams also had arrived at an agreement that they would not push to have an on-site computer as part of the MRP request. Several reasons prompted this change of heart. First, there was currently a great deal of excess capacity in CDS's mainframe computers in Jacksonville. Second, MBS's team felt that they would have their hands full just implementing MRP and did not want to complicate and delay their progress by constructing and managing a computer center. Finally, purchasing a computer would reduce the net benefits expected from investment in MRP, thus endangering its approval to some extent.

After having made their decision, MBS had to wait several months for Nelson to make its decision. During this period, Andrew Gordon kept in continual touch with Hal Messinger and Donald Henderson, always pushing for MRP. Paul Lutchko spent a lot of time talking with Nelson's project team, sharing the information gleaned from MBS's intensive study. Francis Allison maintained contact with CDS's systems and hardware groups:

> I talked to a lot of people in Jacksonville continuously through this process and kept them comfortable so that they would say, "Yeah, we're behind it," those kinds of things, even though they may be scared of it. They feel that we knew what we were doing.... Arthur Brown the [acting] director was the only one who had to sign off. But I kept the guy in charge of management systems, Blake Edwards, informed so that he always felt like we knew what we were doing.... He reports to Arthur. And I always kept people here [at MBS] informed to keep the pressure on Arthur [that] we're doing a good job and those kinds of things. To keep sending the right memos to let him know that people are behind us.

SUCCESS AT LAST (FALL 1985)

In August 1985, the last piece fell in place as Nelson's project team decided to recommend Cullinet as their vendor. Because MBS and Nelson had finally been able to agree on a common system, Messinger was satisfied. The only remaining hurdle was approval of the project by Reston's Executive Committee. Thus the action shifted to much higher levels at Reston as Messinger prepared the senior executives for the coming request. Some of this preparation had already been accomplished at the February meeting in Jacksonville, as well as during Burkhart's trip to Shaler in March. However, at least one board member questioned some of MRP's alleged benefits and Messinger had to spend some time reassuring him.

Messinger, Henderson, and Gordon went before Reston in September with a project request that called for $1.5 million to be spent immediately and $200,000–300,000 yearly for several years afterward. Justification was based on projected increase in market share and profit margins and projected decreases in inventory. Both Gordon and Matthew Faulkner commented on the difficulty in specifying the level of benefits from investment in MRP:

> How do you evaluate what customer service means? We picked a number, X percent margin increase, Y percent market share increase. We

could talk about inventory reductions, those are absolutely real, and I'm convinced that the market share increase is there, but how do you measure that? How do you know that that specific thing created it, or is it better salesmanship or a new salesman or a new product? You just can't, you almost have to have faith. (Gordon)

When you talk about something that's all-encompassing within an organization, as opposed to the more traditional things done in the past...you get rid of three people and you know they're worth $20,000 a year so you save $60,000. You put in this $100,000 machine and in a couple of years, you're fat and happy. Well this isn't the same kind of thing. We can't go out there in the plant and look at MRP machines. And management's got a difficult adjustment to make to invest in this kind of thing, the software. It's an approach to management rather than a piece of hardware. (Faulkner)

Approval was immediately granted by the Executive Committee — the entire process took less than a half hour. Gordon noted that the technical background of many of the committee members contributed to their enthusiasm about the proposed system from a technical as well as general management perspective. Gordon and Henderson telephoned Lutchko and Davis to tell them the good news. More than ten years after Gordon, Benetton, Jennings, and Jefferson had first discussed the possibility of an integrated manufacturing information system, it had finally been approved. In discussing the resolution of the justification process, Gordon made the following comments:

I think being sold on it has to do with the credibility of the person in my chair. I had been with Hal [Messinger] long enough, and I'd had enough successes with other projects...to the point where my credibility was good. So that when I came along and supported and agreed and was enthusiastic about the MRP system, Hal had to listen.... There was no question that he was sold as far as MBS was concerned. He wasn't totally sold that Nelson needed what we needed, but if they [did] they were damn sure going to be going down the same path together.

Paul [Lutchko] had his own series of successes not related to MRP. One of the jobs where he established his reputation and credibility with Hal...was the evaluation of the [Arizona] plant. Everything that you could possibly think about was documented. He was so good at that, there was no question to an analytical person like Hal or me, all having the same engineering background. We could see a staff job so well done, superior to what we had seen in other people to the point where he had established his credibility.

Matthew Faulkner looked back on the process with the following observations:

I think there was a critical decision made several years ago, [but] we didn't realize the importance of it with regard to this whole effort...and that was to locate one of the [CDS] team members here on a full-time basis to be an active team member...not trying to be a member of the team from another location. Because [Francis Allison] got involved on a ground level and worked all the way through it...I think there was no apprehension on the part of CDS any step of the way with regard to how this thing was moving along, and what the pitfalls were. They felt that they were active participants....

Because they were called in at the right time and were involved from the ground up...it would have been a real mistake on our part to...try to do a lot of this work on our own.... It just would not have worked.

I think that the general approach that we took right from the beginning was to try to get as many people involved and educated in this whole thing, in terms of the problems that we were experiencing in trying to grow [and] manage our business. [These problems] were well brought out to the corporate people...to make sure that those folks were familiar with...what was prompting this effort, the problems that we were experiencing, the way that we were approaching the problems, [and] what we thought the ultimate solution was. As a result of that, I think they were more apt to accept our justification, our economics, than perhaps they might ordinarily have been.

In retrospect, this is more in keeping [timewise] with what I had originally expected to occur. Just because of prior experience of working with these other groups, and the immense amount of coordination, and the number of people who had to be involved with the decisionmaking process. I really wasn't surprised that it was taking this long to come to a point where you've got the project submitted, approval, and the final stages of negotiation with a vendor.

PRELIMINARY ANALYSIS

The Monumental Building Supply case covers so much time, and so many actors and events, that it is quite difficult to summarize. It encompasses many of the same themes that were brought out in the other cases, and a few new ones as well. In the early days, progress at MBS depended on the interaction between problems experienced, resources available, and the scope of attempted solutions. Its first attempts at manufacturing systems consisted of small-scale solutions that were implemented with limited resources. As environmental changes caused the problems to increase in severity, it became clear that a large-scale solution, requiring a large commitment of resources, would be necessary.

The discussions about P&MM at MBS brought to light some of the dangers of a large-scale solution; pursuing MRP involved greater risk and expense, more time, and greater neglect of current operating problems. While Lutchko and others were in favor of going for the large-scale solution — MRP — they were acutely aware that by doing so they would be ignoring painful current problems in the factory. Unless a firm is blessed with almost unlimited resources, however, it will have to make concessions by solving today's problems in a limited manner while devising a comprehensive solution that will take some time to implement, thus tolerating today's problems for a while longer.

While MBS's eventual ability to pursue MRP probably reflected some level of corporate confidence in its management team, it is clear that early credibility problems caused delays. The hesitation in accepting MBS's economic predictions for MRP was in part created by its history of cost overruns on data processing projects. However, the disastrous presentation made by Will Gearhart to Messinger and others probably did greatest damage. This presentation set the MRP effort back at least a year and made it a very sore subject within the Building Products Group.

Once the MRP effort actually got started, MBS handled the corporate relations problem quite well. As Matthew Faulkner noted, using Francis Allison as a bridge to CDS was of great benefit. Such an arrangement is perhaps necessary for coordinating a complex project like MRP II. Of course, the importance of the personalities involved should not be underestimated. For whatever reason, Allison was able to get along very well with Lutchko, Faulkner, and other key individuals.

Finally, MBS's difficulty in building a solid consensus with Nelson on MRP played an important role in the case. Given that the firms had different product lines and histories, and each believed it was at least a little better managed than the other, the deck was stacked against consensus. Convincing Messinger of its own need for MRP was the biggest step taken by MBS in this direction. Without pressure from Messinger, it is unlikely that Nelson's investigation would have progressed as fast as it did. Also, MBS's willingness to share information with Nelson's team, to take their needs into consideration, to involve them in the selection process, and to wait for them to decide also helped to overcome the substantial difficulties involved in coordinating the justification process between the two firms.

III FINDINGS AND IMPLICATIONS

9 THE JUSTIFICATION DECISION PROCESS

The previous five chapters presented case studies of organizations that are struggling with the AMT justification decision process. At the end of each chapter, I highlighted the most important aspects of these justification decisions. In this chapter, I will bring together a common set of observations based on all of the cases to construct an overall model of the justification decision process. Particular emphasis is placed on the most common barriers to innovation and how the participants were able to surmount them. I attempt throughout to show how the various elements of justification decisions relate to one another to form an integrated process.

The overall model argues that innovation champions need to satisfy three types of criteria in order to have an initiative approved: strategic/financial, interpersonal, and political. The decision process requires, in large part, that these individuals struggle to meet these criteria even though the organization is changing around them.

THE STRUCTURE

The basic structure of AMT justification decisions derives from their place within the capital budgeting framework. Requests to expend capital come from lower levels in the organization and must be approved by managers at successively higher levels. If a manager at any level turns down a request, it does not progress to the next level. Requests for larger amounts of capital require higher levels of approval, although organizations obviously differ on the amount of discretionary spending allowed at various levels.

The participants in the innovation decision process for AMT can be divided into four categories. Approval for the process has to pass through each of these levels. First, there are the lower level technical personnel responsible for initiating the process. These individuals are often just out of college and are fascinated by advanced technology's potential for improving operations. In the firms I studied, technical personnel had such titles as design engineer (e.g., Gene Jackson at Defense Technology) or manager of production planning. They often learn about AMT by reading trade magazines or attending conferences.

AMT investigation can be instigated either by an ongoing operational problem or simply by the awareness of a new technology. In three of the cases, operational problems generated a search for solutions, which ultimately resulted in a recommendation that AMT be used. Temple needed to lower its labor costs in order to be globally competitive. American Plumbing Fixtures had a plant that was falling apart; its problems were compounded by noncompetitive labor costs. Monumental Building Supply was finding it very difficult to control the multiplicity of products in its plants and was having problems meeting its deadlines. At both International Metals and Defense Technology, however, there was no pressing problem for which AMT was the solution. Awareness of a new technology initiated the justification process in these two firms.

The second level of approval in the process takes place at the technical management level. These individuals have the responsibility for balancing technical and business factors (Maidique 1980). In the firms I studied, these individuals held such titles as director of technical services and vice president of operations. Pete Jordan of Temple Laboratories exemplifies the role of technical management. In some cases, these individuals initiated the proposals.

The third (and, in some cases, final) level of approval comes from the company's management, usually from the president. These individuals, who establish the strategic directions and organizational constraints within which lower level managers must work, are more financially oriented than are the technical managers. They consider the potential adoption of technology from the standpoint of company strategy and long-term competitiveness. In companies that are not part of a corporate structure, these managers have the final say as to whether funds will be committed to advanced technology. At Defense Technology, for example, the process ended with approval from Bruce Kennedy. At Monumental Building Supply, however, Andrew Gordon had a lot more work to do after giving his approval.

In organizations that are divisions or wholly owned subsidiaries of corporations, one or more additional levels of approval are involved. Managers at the corporate level are usually even less technically and more financially oriented than the company managers, and they know quite a bit less about the division's or subsidiary's specific business conditions. They see the adoption of AMT as simply an investment to be compared with alternatives in terms of risk and expected return. The tenor of American Plumbing Fixtures's presentation to Diversified Corporation illustrates the emphasis placed on the financial perspective at the corporate management level.

Formal mechanisms exist at all levels for passing recommendations up the line. These mechanisms consist primarily of standard forms (usually called "authorization requests") that require signatures from each approving manager, as well as of formal presentations. As we will see, these formal mechanisms are almost always augmented by more informal methods of communication.

THE PROCESS

The dynamics of the justification decision process consist essentially of the lower level participants' (proponents, or "champions") attempts to convince the upper level participants to approve the project. Thus the term *justification,* which conveys the idea that one set of people has made a decision and is now attempting to justify this decision to those who have the final say.

At any given time, the participants can be divided into two groups: those who already have been convinced and have given at least verbal approval, and those whose approval is still being sought. As one might imagine, lower level managers usually are the first to be "sold," and the highest level managers are last. (This terminology of *selling* the technology was consistently used by the people in the firms I studied.)

In the early stages of the process, the middle managers are skeptical, critical thinkers who challenge the assumptions and analysis of the proponents. Once sold, however, they themselves become champions and devote enormous energy to selling the technology to the next level up. While it seems probable that some managers would be only marginally supportive of a project and approve it without helping to sell it further, they know that a lukewarm endorsement can kill a project and therefore either turn it down or become boosters.

AMT justification decisions often have a strong emotional or even passionate tone. Recent research in social cognition has been devoted to the emotional character of individual preferences (Park, Sims, and Motowidlo 1986). With individual reputations, to say nothing of company futures, on the line, emotion clearly plays a strong role in the decision process.

Tom Kidwell at International Metals described his dissatisfaction with current practices that led him to support CIM as "an aching inside of me." People at Monumental Building Supply conveyed their "warm feeling about MRP." Andrew Gordon, after hearing about a memo that questioned the merits of MBS's approach to MRP, began a phone conversation with the words, "I'm livid!" At Defense Technology, a prolonged process finally resulted in the approval of a CAD system; when one of the initiators heard the news, tears streamed down his cheeks. Clearly, the proponents of AMT can become very emotionally involved with their proposed innovations. Given this emotional commitment, it is not surprising that they will do whatever it takes to get the projects approved.

AMT APPROVAL COMPONENTS

I have used the term components rather than criteria because it suggests that something is being *constructed*. The proponents are the "architects" of AMT approval. As I have indicated, the number and seniority of the architects increases as the process evolves. The components necessary to complete the structure (i.e., obtain approval) are threefold: strategic/financial, interpersonal, and political. Although many combinations of the three can be successful, some minimal level of each is probably needed. AMT proponents must skillfully build these components into a structure for approval that will succeed in their organization.

The strategic/financial component of AMT approval requires that the proponents demonstrate that the proposed investment will improve the firm's financial or competitive position. The interpersonal component involves the track records of the champions and their commitment to the project. Finally, the political component stems from the fact that, in addition to those individuals who have formal veto power over the project, the support of any number of other organizational actors may be necessary. If these factors are not present to some degree, approval for the project either will be delayed or withheld completely.

The Strategic/Financial Component

In the capital budgeting framework within which AMT decisions are made, the expected value of an investment is based on risk and expected return. As I mentioned above, strategic and financial considerations increasingly dominate technical concerns as the decision reaches higher levels. Calculating a level of return for most AMT projects is an extremely difficult undertaking (e.g., Kaplan 1986).

AMT proponents need to demonstrate the financial justification for their proposals by using such measures as internal rate of return, payback period, and net present value. These techniques, and their assumptions, are fairly complex (Gold 1983) and involve art as well as science. Thus the financial justification process takes place in an atmosphere of profound uncertainty: no one knows exactly what the return will be. Even if the technology is adopted, it is often impossible to know whether the expected level of return has been achieved, due to changes in production rates, competitor response, and the like.

How the proponents approach the strategic/financial component depends on the difficulty they encounter in demonstrating financial benefits that exceed the corporate "hurdle rate." This is the rate of return that is necessary for an investment to be approved; it ostensibly reflects the firm's cost of capital. If such a return can be demonstrated unambiguously, the proponents' job is easier. This return is often based on reducing labor costs, the traditional method for justifying investments in automation. At Temple and American Plumbing Fixtures, the only cases in which the financial component was not an issue, the projected returns were based almost entirely on labor cost reductions. In most industries, however, labor represents only 15 to 20 percent of total product cost, making it difficult to justify automation based primarily on labor cost reduction.

If the expected return does not quite meet the hurdle rate, or is based on such "questionable" benefits as reduced inventory or increased market share, the process of securing the strategic/financial component becomes more complicated. The most common approach is for the proponents to exaggerate the benefits sufficiently to meet the hurdle rate (Bower 1970). As Pete Jordan at Temple put it, "If it's within 5 percent of the hurdle rate, I lie." Of course, there are times when the expected return is clearly unacceptable: these projects usually do not proceed beyond the technical management level.

Depth. Perhaps due to the uncertainty surrounding the calculation of the benefits of AMT investment, senior managers sometimes assess the merits of a proposal by assessing the depth of its analysis. Although senior managers cannot possibly delve into all the details of the analysis, they expect the proponents to have done so. They also periodically check the depth, or thoroughness, of the analysis by asking specific detailed questions and assessing both the answer's thoroughness and the confidence with which it is delivered. If the proponents stumble over too many of these questions, the decision process may be abruptly terminated.

Another approach taken by the proponents is to augment the financial analysis with a rationale based on strategy. This tactic becomes more prevalent as the benefits of the AMT investment become murkier, or as the cost or technical risk increases. These arguments express themes that include the expectation that competitors will be adopting this technology, customers' perceptions of the firm will be enhanced, this is how things will be done in the future, and so on. As Carter (1971) has noted in another context, proponents will try to tailor their arguments to address issues that are important to senior decisionmakers. This contingency plan of stressing "intangible" as well as financial benefits for AMT was expressed by Matthew Faulkner at Monumental Building Supply:

> The benefits of MRP that we've been looking at are so substantial that, unless the development costs change appreciably, I think the economic justification is there on the basis of our traditional payback and ROI criteria. However, if it becomes a marginal situation, we would have to dig in a little deeper and try to look for some of the other benefits. Not the easily identifiable dollars-and-cents benefits, but you know, long term that this is the thing you should do.... Smart businesses have recognized MRP as not so much a capital investment, but more as an approach, as a new environment.... If the economics of MRP were not clear-cut enough, I think we would have to promote that concept about the new environment.

As AMT's projected benefits become less quantifiable, emphasis shifts from documenting (and exaggerating) economic benefits to this type of future-oriented, strategic rationale. In the extreme case, financial projections may not be prepared at all and the proponents stake their case solely on their ability to construct a strategic rationale (sometimes called a "story") that supports investment in AMT. This approach is probably rare, but it characterizes Defense Technology's justification of CAD. As Keith Thompson put it:

> We would have attempted to do a study where you could come up with numbers and you could say your payback is this. But because there are so many estimates, it becomes nonsense. There's no point in trying to do that. You can make of it whatever you want with an estimate. Whatever assumptions you make, you come out with a new answer.

Needless to say, such an approach to justification does not come easily. Ron Jenkins, Deftech's vice president of manufacturing, described the preparation necessary to bring such an unusual authorization request to senior management:

> I think that we planted that seed way back when we first talked about it and put it into the capital budget. We said, "We're not sure that we can show this in the typical sense of payback that you're used to envisioning. You may have to consider other intangibles and other benefits that can come out of this. You are going to have to look at this and make some judgments."

When the request was submitted, no financial analysis was presented. The rationale for this was stated as follows:

> Traditional cost justification techniques are inappropriate for this type of purchase. Benefits are generally long range and do not lend themselves to a "present vs. proposed" method of analysis.... As outlined above, CAD will allow us to...do some things we presently cannot do. We are dealing with quality and content of output.

Translation. Regardless of the approach taken to satisfy the strategic/financial component, some level of "translation" will be necessary. Translation is the process by which the technical managers convert the excitement and enthusiasm of their personnel into a rationale that can be understood and accepted by senior managers. Maidique (1980) and Burgelman (1983) have noted similar middle management roles found in the technological innovations and new ventures, respectively. George Coyle, director of technical services at Deftech, described how the process worked:

We asked each user group to draft what they wanted to say. We did most of the rewriting because we know what needs to be said so people will approve it.... Technical people tend to become too technical, so we put it in layman's language. It was a translation. This was extremely important. We want people from the administrative and financial disciplines to read it and come to the same conclusion. It's difficult for some of them to understand all the technical jargon, so we put it in as easy terms as possible.

In summary, one of the components crucial to AMT justification is strategic/financial. The tactics used to secure approval on this basis vary; they depend on the level of difficulty involved in demonstrating an acceptable financial return. When the projected return is clearly acceptable, proponents let the numbers speak for themselves. If the expected return is clearly unacceptable, the project is not forwarded to higher levels. In marginal situations, the benefits are exaggerated by proponents in order to hit the company's hurdle rate. In many cases, however, sufficient ambiguity exists that the numbers need to be bolstered by a story about strategic and intangible benefits. The emphasis on this latter tactic increases as the financial returns become smaller and/or harder to calculate, or as the cost or risk of the project increases.

Senior managers use the depth of the proponents' preparation as an indicator of the quality of the proposed investment. One of the tactics by which the proposition is conveyed to senior managers is translation, whereby technical managers recast their departments' technology-based rationale for adoption into themes and concepts that will be appreciated and accepted by senior managers.

The Interpersonal Component

If business organizations were in fact the purely economic, rational entities envisioned by classical microeconomic theory, then perhaps demonstrating the strategic/financial benefits of AMT would be the sole component of justification decisions. However, given the organizational context in which such decisions are made, and the indeterminate nature of the financial analysis that informs them, other components become important to the process. The interpersonal component involves the need for AMT proponents to demonstrate two characteristics: credibility and commitment.

Credibility.   A number of observers have noted that when faced with uncertain decisions, managers rely on their perceptions of the individuals who are proposing the courses of action rather than on the proposals per se (Bower 1970; Carter 1971; Lyles and Mitroff 1980; Burgelman 1983). Discussions often center around the credibility, or "track record," of the champion. Credibility emerged as a very strong theme in the AMT cases that I studied.

Credibility is a crucial asset among organizational actors. As Carter (1971) has noted, maintaining credibility limits the possibility of biasing information, which distorts communication in decisionmaking (Cyert and March 1963). Credibility stems from having "delivered" in the past — having lived up to one's promises. It is also a relationship-specific quality; one may have built up a great deal of credibility with one's old boss, but a new boss requires one to rebuild credibility.

Credibility implies not only that a proponent's projections and promises are reasonable, but that he or she possesses the managerial skill to make them come true, even under trying circumstances. In large organizations where managers cannot possibly confirm even a fraction of the claims made to them, it is not surprising that credibility is a highly valued and jealously guarded asset. One statement of the importance of credibility is provided by Fred Barnes, vice president of operations at American Plumbing Fixtures and chief proponent of its new robotic system:

> You have to be able to look at the past and say what you have done.... What is your success rate? That is very crucial to getting anything approved.... If you have a good track record and you do your homework properly, then you have a chance for success.
>
> My boss was not a difficult sale. Some conversation and additional analysis...Then again, I had the credibility. That's why I said [what I said] about your past record being so important. I had worked for him for almost four years. If I had been a brand new man on the block, it would have been a lot tougher sale.

The issue of crediblity also figured prominently in Monumental Building Supply's choice of a consultant to help in selecting an MRP system. While MBS's managers believed that one consultant was superior on the basis of technical knowledge and experience, the other consultant was chosen. Because the consultant was employed by Reston's (MBS's parent) auditor, the proponents felt that he would have greater credibility with the corporate officers when their approval was being sought.

Commitment.   The interpersonal component also involves commitment. Commitment is multifaceted; it consists of certainty, responsibility, and enthusiasm. Senior managers are very reluctant to approve risky projects — especially those with big price tags. This perception of risk is overcome by the proponents' promise that the innovation will work and their willingness to take responsibility for making it work.

This practice obviously increases the risk to the proponents, who are seldom as sure as they seem. They are faced with a trade-off between credibility and certainty. In order to gain approval, they know they must be very confident in their discussions with management; yet if the innovation is a failure, their credibility is destroyed. On the other hand, if they present a more realistic picture of the risks involved, thus maintaining their credibility in case of failure, their senior managers may simply kill the project. The personal risk involved in selling an unproven technology to management was dramatically illustrated by one manager's words: "We put our heads on the block and said it would work."

The responsibility factor of commitment is exemplified by events involving Deftech's approval of CAD. As Ron Jenkins, vice president of manufacturing, tells it:

> This thing had been viewed all along as a joint venture between engineering and manufacturing. It came down to the final signatures on the capital appropriations form, and there's one line on there that says cognizant vice president, the guy that's supposed to be responsible for the thing. When it came down to signing, it was between myself and our vice president for research and development, and I said, "We'll both sign it." But the executive vice president said, "I think there should be only one signature on there. . . . If manufacturing isn't willing to take this on and make damn sure they they use it, then you're not going to get the system." So I signed.

Enthusiasm is the third aspect of commitment that AMT proponents need to demonstrate. While credibility can only be developed over the long run, commitment, especially enthusiasm and certainty, can be demonstrated in a shorter period of time through a practice that might best be called "softening." Softening consists of repeatedly mentioning the tremendous potential of the proposed technology, the imminency of the authorization request, and enthusiasm at the prospect of having the technology. The amount of softening seems to increase in proportion to the scope, expense, or risk of the proposed innovation. As Ken Tornatski at Temple Laboratories put it:

For a project of this magnitude, you don't just write an AR [authorization request] and sent it to [corporate headquarters]. There is sort of an initial indoctrination, so it is not like just throwing it at them—"There it is, we want it." This way, they have already been exposed to it. There is some familiarization.

Charles Bennett, manager of manufacturing engineering at Defense Technology, provided another example of how the softening process works:

I think being enthusiastically behind it, being a booster is what [makes a difference]. . . . It's not sold in the meetings, with the executive vice president sitting there reading it. It's sold by. . . running into someone at the coffee machine, and they say, "How's that CAD coming?. . . Do you really need it?" "Oh, absolutely. We've got to do it." That's [what] sells this thing. The meeting formalizes it. . . . I really believe that you get a lot of this stuff over by being 100 percent positive in the informal kinds of contacts.

In summary, the interpersonal component is comprised of the personal characteristics of credibility and commitment that are necessary for project approval. Credibility is based on one's prior successes, while commitment involves certainty, responsibility, and enthusiasm. In their presentations to management, proponents make trade-offs between credibility and certainty. While it is difficult to enhance one's credibility in the short term, others with credibility can be brought into the process. Commitment can be demonstrated to some extent through softening, as well as by the willingness to take responsibility.

The Political Component

The political component is the third and final piece necessary for project approval. The term "political" is used here in a somewhat narrower sense than usual; it refers to the necessity of securing support for AMT among organizational actors who are neither proponents nor final decisionmakers. This support must be in place for approval. Therefore, proponents generally address the political component before making serious attempts to secure approval from above.

The political component is important for much the same reason as is the interpersonal component. Because the analysis supporting investment is usually indeterminate, other means must be found for making decisions. Social psychologists have long been aware that consensus among a group of people is a powerful force in forming beliefs and attitudes, particularly under ambiguous circumstances (Asch 1951; Sherif 1936). Pfeffer (1981) has also noted that consensus within organizational subunits leads to organizational power.

Solidarity.   The particular challenge for AMT proponents stems from the fact that support needs to be gleaned from an unusually diverse set of individuals, especially for such software-intensive technologies as CIM. Because these technologies reach into many corners of the organization, senior managers use the level of support across the breadth of the organization as a gauge in making their decisions. If the support is not broad and deep enough, the proposal will probably be thwarted before reaching the highest levels. I have termed this combination of consensus and commitment necessary for approval of risky ventures "solidarity."

Solidarity was a particularly crucial issue for International Metals and Monumental Building Supply. IMI was attempting to create a corporate-wide CIM initiative. MBS's corporate management had insisted that the firm embark jointly on its MRP project with their sister company. These two cases demonstrate both the importance and the difficulty of building support for AMT among a diverse set of actors.

In order to create CIM at International Metals, the support of at least four different organizational constituencies was needed: engineering, research and development, management information systems, and the business units. Helen Evans, who had spent months helping to set up a meeting for key representatives of these constituencies, commented on her feelings after the meeting:

> I was a little frustrated at the end because the strategy they came out with, the CIM strategy, was what I had expected going in. [My boss] said, "Wait a minute, there's a difference between you having made up your mind as a result of having spent four months [working on the issue] and the corporation forming consensus and commitment around [CIM]. . . . The meeting was to develop the commitments and working relationships to do something about it." And out of everything that happened at the meeting, the most valuable thing is that these guys closed ranks.

Subsequent events put the solidarity that evolved from that meeting to a severe test. When the CIM strategy was presented to corporate management, the proponents were literally thrown out. But they were able to regroup and make a second presentation that was ultimately successful. As Tom Kidwell put it:

We were really getting worked over, but they noticed that we never broke. I'm sure of that.... When the corporation started coming back at us and saying why are you doing this, we held tight and convinced the top of the corporation that CIM is critical and we have to do something about it.... We believed in CIM enough...to hang tough.

At MBS, enormous effort was devoted to "converting" a recalcitrant sister company to the "gospel" of MRP. By corporate edict, the MRP project was to be a joint effort between the two companies. Each company had been acquired by the corporation at about the same time, and each felt that it was a little better than the other. Developing solidarity in this environment was a Herculean task, which consumed almost two years of the decisionmaking process. Paul Lutchko's comments are representative. Here is his reaction to a meeting with a top manager at Nelson, the sister company:

We wanted to at least set some groundwork up so that we could present some agreed-upon approach to this joint effort. We didn't even get past talking about the differences in scope for the two companies.... I look at the joint study and I say, if a convoy can only go as fast as its slowest boat, it worries me.... Nelson is still reluctant to pursue the path that we have chosen.... They feel that MRP is a buzzword for us, and that we're cramming it down their throats.... It's going to be a real dogfight trying to reach that common point where we can stand back and discuss what we have in common and what we don't have in common.

The tactics used by the proponents to build solidarity among the subunits were as varied as the organizational scenarios they faced. At International Metals, outside experts were brought in to deliver the message about CIM so that it would not be perceived as coming from one of the parties with a stake in the outcome. The company also employed technologically sophisticated process facilitators who used structured techniques for consensus-building. At Defense Technology, engineering tried to gain approval for CAD without prior involvement from manufacturing; it met with the predictable failure. A subsequent attempt to go over the heads of manufacturing also failed. The groups were united only when engineering took manufacturing's needs seriously.

Similarly, the situation at MBS was resolved only after many months of sharing information, polite inquiries, working with those at Nelson thought to be the most sympathetic, and a good deal of waiting. While the delays involved were frustrating to the proponents at MBS, the ultimate success of the project was due in no small part to the time taken to construct a common understanding about MRP.

Visualization.   One tactic used to build political support derives from the abstract nature of many of the technologies that comprise AMT, particularly CIM. In order to gain support from those whose support is necessary, proponents must find some way to make the technology tangible to them. In a sense, visualization is a variant of translation, but it is visual rather than linguistic in nature, and it attempts to convey an idea or vision, rather than a financial benefit.

At International Metals, CIM was brought to life for one of the business unit managers when he was taken to see a European plant where CIM had been implemented to some degree. As Patrick Broadbent put it, "It was a stroke of genius to take [the trip]. Now when we tell people about what we saw there, they don't say, 'That's just you guys talking'. They say, 'You're right'." Bruce Lindsay from MIS was very concerned about the corporate-wide visualization problem and planned to help solve it by using a CIM demonstration:

> For people who have never been around computing, you're asking too much to ask them to buy into [CIM]. All it is is words. They're going to have to see it and touch it. Once they've done that, I think they become extremely strong supporters. . . . I'm going to use [the CIM demonstration site] as my showcase to bring people in to look at it, touch it, feel it, walk around it and say, "Yes, now I can see what you're saying."

In addition to plant tours and demonstration projects, computer simulations are another common way of achieving visualization. At Temple Laboratories, an animated computer simulation was instrumental in convincing a number of people that the robotic cell concept would work. While their methods differed, all of the AMT proponents found some way to make their ideas tangible to the managers who needed to feel comfortable with them.

In summary, due to both the uncertain nature of analysis concerning AMT investment and the diverse group of people who will often be affected by it, senior managers assess the solidarity behind an idea as an index of its worth. Proponents must therefore sell their ideas to people who are not in the formal chain of approval. Strategies used to accomplish this include sharing information, waiting patiently for another group to accept an idea, and using various techniques for visualization.

THE ORGANIZATIONAL CONTEXT

In the preceding sections, I have described the three components of AMT justification that must be put in place in order for proponents to be successful in their innovation attempts. This process obviously does not take place in a vacuum. AMT justification is just one stream of activity among many in an organization of any size. In this section, I will discuss some of the ways in which the organizational context of these decisions affects the decisionmaking process.

Distraction of Current Business

It is clear that constructing AMT approval is an enormously time-consuming process. Performing financial analyses, softening senior managers, and selling across the organization all take time. Freeing themselves from their ordinary responsibilities in order to work on the AMT project is a major challenge faced by proponents.

George Coyle at Deftech said, "We didn't look very hard [for a CAD system] because we were so busy." Paul Lutchko at MBS was frustrated because "the MRP project always takes a back seat to the daily fires." Releasing four individuals from their daily responsibilities to investigate MRP was a major milestone in this case. Sanford Turner at International Metals also noted the difficulty of escaping from day-to-day obligations to think about the technology of the future:

> In an operating entity, the problem is today's business, and that's where you gravitate all the time. You have to get off and think about new and innovative things, which is very difficult. So we were looking for approval of a direction that would allow us to go off and worry about CIM [and] keep ourselves out of the mainstream and the daily problems.

Structural Location of Proponents

The location of the AMT proponents in the organizational structure also plays a role in the decisionmaking process. Simply put, the more highly placed in the organization the proponents are, the easier it is for them to get what they want. While much has been written about the structural and personal characteristics that lead to power in organizations, the dominant source of power in the cases I studied was simply the formal authority of hierarchical position.

Organizations that have developed some commitment to advanced technology often formalize this commitment by elevating one or more individuals to higher levels in the organization. At International Metals, Sanford Turner was chosen to spearhead the CIM effort and was simultaneously moved up a level, thus symbolizing the importance of the initiative. The position of CIM leader eventually was broadened, and Turner's successor in this position was given the same title as the business unit managers.

Apart from simple hierarchical level, placement in the formal chain of approval is also a source of power in AMT justification (Pfeffer 1981). This is illustrated by events in Deftech's search for a CAD system. When the earliest CAD proponents in the design area had identified a system they wanted to purchase, they contacted the manufacturing group. Since manufacturing did not feel that their needs were met by this system, approval was withheld. Even though design and manufacturing were on the same hierarchical level, manufacturing was able to force design to look for another system because it — not design — was part of the formal chain of approval.

Entangled Issues

Cohen, March, and Olsen (1972) noted the tendency of problems or issues to become entangled with decisions in organizations. They also pointed out that most decisions get made when such problems either go unnoticed or are temporarily attached to another decision. Perhaps because of the new and ambiguous nature of the AMT initiatives in the firms I studied, I found that tangential issues often became attached, usually resulting in the confounding and slowing of the approval process.

The major problem attached to the MRP decision at MBS was the decentralization of computing hardware. MBS's parent company had traditionally operated in a centralized data processing environment. Several people at MBS thought that MRP implementation might provide an opportunity to break with that tradition — they wanted a mainframe computer installed locally. The temptation to make MRP "the battle flag for decentralization," as Paul Lutchko put it, was increased when the corporate data processing manager was given early retirement, supposedly due to his resistance to decentralization. Eventually, however, the MRP proponents decided to divorce the two issues, noting among other things the extremely political nature of decentralization.

Decentralization was also an issue for International Metals. In this case however, it was an issue of people rather than machines. One of the reasons for the CIM proponents' initial dramatic failure with senior management involved the entangling of the decentralization issue with CIM. The company had recently undergone a substantial decentralization and reduction of technical personnel, and managers who heard the CIM presentation felt that the proponents were advocating recentralization. Tom Kidwell attests:

> What had happened is that the subject of decentralization had gotten confused with the subject of networking and architecture. Those of us who [were pushing for CIM] had never dealt with the question of whether computers should be used in a centralized or decentralized corporation.... but when we used words like architecture and networking, people thought that meant centralization at a time when they were trying to be decentralized.

Proximity

Another contextual dimension that can affect the AMT justification decision process is the proximity of the players. At Deftech, much was made of the fact that all of the players were in the same building, which facilitated the informal conversations that move the decision process along.

This was atypical, however, for the cases I studied. MBS, for example, is located over a thousand miles away from both its sister company and the central data processing group, and several hundred miles away from corporate headquarters. Such distances are not unusual, since location decisions are made for reasons other than to facilitate decisionmaking processes. Major strides were made at MBS when a systems analyst from the central computer group was placed with the company on a full-time basis. This assignment greatly aided in the coordination between MBS and the central computer group.

Unrelated Events

In almost any decision process, and certainly in those I studied, events in and around the organization will affect the nature of the process.

Mintzberg, Raisinghani, and Theoret (1976) call these events "interrupts." In a similar vein, Bower (1970) discusses the "situational context." The occurrence of these events, and the effects they will have on the decisionmaking process, are usually unpredictable. They include personnel turnovers, sudden illnesses, gross increases or decreases in demand for a product, and the availability of a new and better version of a particular technology.

When a key manager at Temple was promoted, his replacement delayed the AMT decision process for several months. When MBS's senior management decided to open a plant in a new geographical region, Paul Lutchko was pulled off the MRP team to do the financial analysis. (Ironically, while this delayed the MRP decision for some time, Lutchko's superior performance on this task sufficiently enhanced his credibility that selling MRP up the line was made much easier.) American Plumbingware's parent company was acquired by another corporation during the process, resulting in a new and different approach to justification. The decisionmaking process in this same case was accelerated when a recalcitrant manager retired.

While it is difficult to predict the timing or effects of these unrelated events, one can be reasonably sure that they will occur in any decision process of reasonable duration. An AMT justification decision that flowed smoothly from start to finish would be the exception rather than the rule.

SUMMARY AND CONCLUSION

I have described the nature of the decision process for AMT justification based on my experiences in five companies. The model that emerges from these cases can be summarized in the following set of propositions.

1. The major participants in AMT decisions are technical personnel, technical managers, company or division managers, and corporate managers.
2. As the AMT proposal progresses up the chain of approval, participants change roles from critical decisionmakers to innovation proponents.
3. Innovation proponents become very emotionally involved with the projects they champion.
4. Decisionmakers' assessments of innovations are partially based on the innovations' financial and strategic characteristics.

5. When the projected benefits of AMT fall slightly short of the firm's hurdle rate of return, proponents often exaggerate the benefits enough to meet this rate.
6. As the cost, risk, or technical uncertainty of an AMT project increases, proponents are more likely to augment the financial analysis with a rationale based on intangible benefits.
7. The success of proponents in convincing decisionmakers depends partially on their ability to translate their perceived benefits of AMT into benefits that senior managers can appreciate.
8. Decisionmakers' assessments of innovations are partially based on their perception of the credibility and commitment (certainty, responsibility, and enthusiasm) of the proponents.
9. Proponents try to demonstrate commitment to the innovation by "softening" decisionmakers, a process that becomes more common as the cost, risk, or technical uncertainty of the proposed AMT increases.
10. Decisionmakers' assessments of innovations are partially based on the degree of support that the innovations receive from key individuals outside the formal chain of approval.
11. Proponents attempt to develop support from individuals outside the formal chain of approval before approaching decisionmakers. They are successful in gaining approval to the extent that they are able to build solidarity (consensus and commitment) among the outside individuals.
12. In order to build solidarity, proponents must find a way to visualize the proposed technology for decisionmakers.
13. Proponents are more successful in their efforts when they are freed from day-to-day responsibilities, when they are higher in the organization and more central to the decisionmaking process, and when they are located in close physical proximity to decisionmakers.
14. Organizational contextual factors such as attached issues and unrelated events have pervasive but unpredictable effects on innovation decisions.

REFERENCES

Asch, S.E. 1951. "Effects of Group Pressure upon the Modification and Distortion of Judgments." In *Groups, Leadership, and Man,* edited by H. Guetzgow, pp. 177–90. Pittsburgh: Carnegie Press.

Bower, J.L. 1970. *Managing the Resource Allocation Process.* Boston: Graduate School of Business Administration, Harvard University.

Burgelman, R.A. 1983. "A Process Model of Internal Corporate Venturing in the Diversified Major Firm." *Administrative Science Quarterly* 28, no. 2 (June): 223-44.

Carter, E.E. 1971. "The Behavioral Theory of the Firm and Top-Level Corporate Decision." *Administrative Science Quarterly* 16, no. 4 (December): 413-29.

Cohen, M.D.; J.G. March; and J.P. Olsen. 1972. "A Garbage Can Model of Organizational Choice." *Administrative Science Quarterly* 17, no. 1 (March): 1-24.

Cyert, R.M., and J.G. March. 1963. *A Behavioral Theory of the Firm.* Englewood Cliffs, N.J.: Prentice-Hall.

Gold, B. 1983. "Strengthening Managerial Approaches to Improving Technological Capabilities." *Strategic Management Journal* 4, no. 3 (July/September): 209-20.

Kaplan, R.S. 1986. "Must CIM be Justified by Faith Alone?" *Harvard Business Review* 64, no. 2 (March/April): 87-95.

Lyles, M.A., and I.I. Mitroff. 1980. "Organizational Problem Formulation: an Empirical Study." *Administrative Science Quarterly* 25, no. 1 (March): 102-19.

Maidique, M.A. 1980. "Entrepreneurs, Champions, and Technological Innovation." *Sloan Management Review* 21, no. 2 (Winter): 59-76.

Mintzberg, H.; D. Raisinghani; and A. Theoret. 1976. "The Structure of 'Unstructured' Decision Processes." *Administrative Science Quarterly* 21, no. 2 (June): 246-75.

Park, O.S.; H.J. Sims, Jr.; and S.J. Motowidlo. 1986. "Affect in Organizations: How Feelings and Emotions Influence Managerial Judgment." In *The Thinking Organization,* edited by H.P. Sims, Jr., and D.A. Gioia, pp. 215-37. San Francisco: Jossey-Bass.

Pfeffer, J. 1981. *Power in Organizations.* Boston: Pitman Publishing.

Sherif, M. 1936. *The Psychology of Social Norms.* New York: Harper and Row.

10 IMPLICATIONS FOR MANAGEMENT AND RESEARCH

In Chapter 1, I described how the problem of AMT justification relates to manufacturing firms' global competitive environment. In Chapters 2 and 3, I described some of the existing studies that have been done of capital budgeting and inovation decisions, as well as my AMT justification study. Chapters 4 through 8 presented case studies of how AMT was justified in five organizations, and Chapter 9 summarized my findings from these cases in a model of the AMT justification process. This final chapter will describe the implications of my findings for managing the justification process and doing future research on innovation decisions.

IMPLICATIONS FOR MANAGEMENT

The most basic finding of my research is that a firm's ability to justify AMT depends largely on the strategies and activities of the proponents. The difference between one firm that is able to achieve its strategic objectives through AMT implementation and another firm that gradually withers away due to outdated technology may lie in the interpersonal and organizational dynamics that comprise the justification process.

The success of AMT proponents may be reflected not only in whether they are ultimately able to gain approval for their project, but also in how long it takes to do so. Ineffective championing often results in delays in scheduling meetings, requests for more detailed analysis, and so on. As one of the engineers at Temple Laboaratories put it, "Sometimes it drags on because [senior managers] really don't want to do [the project]." While these delays may be eventually overcome (both Defense Technology and Monumental Building Supply recovered from long delays), they are not without cost. If an investment in technology is projected to return, say, $50,000 a year, this opportunity is forgone while the project awaits the necessary signatures.

More ominously, delays allow competitors to implement technology first, thus gaining whatever advantages accrue to the "first mover." Delaying justification may mean the difference between the opportunity to gain a strategic advantage over the competition and trying merely to regain ground that has been lost. Paul Lutchko, the MRP project leader at MBS, felt that his company could have reaped much greater strategic benefits from MRP had it been implemented several years earlier.

The Value of the Process

At first glance, it might appear that the dependence of AMT decisions on the type of interpersonal and political dynamics I have described represents a failure of management. (At a minimum, these findings certainly contradict the popular stereotype of business decisions that are based solely on the bottom line.) I feel, however, that for the most part, the justification decision process as it is practiced is not capricious and, in fact, is quite functional for making this type of decision.

My findings do not imply that managers are choosing to ignore hard data in favor of focusing on more subjective considerations. Managers in all the companies I studied made every attempt to take advantage of what hard evidence existed. The problem is that there are simply no "right answers" for such decisions — at least none that can be specified in advance. As the CAD proponents at Defense Technology pointed out, your answers depend on what assumptions you make.

In this environment, a proponent's track record and reputation may be as good an indicator as any of whether an AMT initiative is a good investment. If an individual has succeeded in other areas in the past, it is not unreasonable to assume that he or she will continue to succeed in this new area. Using a proponent's level of commitment to a project also helps a manager to make a decision. Commitment indicates both the quality of the project and the likelihood that the proponent will be driven to do whatever it takes to make it a success.

A similar argument can be made for reliance on solidarity: the existence of a committed group of people indicates both that the idea has some broad appeal and that the support crucial to implementation is already in place. In short, senior managers should rely on such indicators as credibility, commitment, and solidarity.

Activities such as translation and visualization can also help organizations make effective decisions about technology. Both of these activities are directed at making the new technology concrete and meaningful to senior managers. In that they at least partially redress the imbalance between the obvious risks and sometimes abstract benefits of advanced technology, they are a mechanism for giving the proposed technology fair consideration. Decisions should certainly be based on a full appreciation of the potential of AMT. Whatever tactics may be used to create this appreciation can only help an organization make better decisions.

In other words, rather than trying to tighten or depoliticize the justification process, managers should recognize that the process as it has evolved is a reasonable one for making important decisions about uncertain technology. While the process may appear messy and chaotic, it can, in the long run, help to produce good decisions under difficult circumstances.

Recommendations

How can managers use this picture of the justification decision process in their attempts to justify AMT in their own firms? While organization-specific circumstances may vary the process, the following general guidelines can improve an individual's chances of securing approval for advanced technology. Needless to say, all of these recommendations are grounded in the belief that organizations need to implement AMT more extensively than has been done to date. If American manufacturing firms are to remain competitive in world markets, they will need to successfully utilize these advanced technologies.

First, do your homework. All potential benefits of AMT should be investigated. These include not only labor cost reduction but also improvements in inventory, materials, warranty costs, space, market share increases, quality, and flexibility. If any of these benefits apply, include them, but be sure you can defend your authorization request. Make certain that you are convinced that the project is worthwhile. If you have any doubts, you should reconsider its validity before trying to convince anyone else.

Assess the proposed AMT project for its consistency with your organization's strategy. Even a project that has a good return might not be worth pursuing if it does not match your firm's strategy. If your firm is striving to produce the highest quality product in its industry, a project that is aimed at cost reduction is not as good as one that is directed toward maintaining machining tolerances. If the project does match with the company's strategy, translate its technical benefits into strategic ones.

Realistically assess whether you have the credibility to gain approval for the product. The amount of credibility necessary will depend on how obvious the project's benefits are. Both your track record and the amount of time you have worked for your current boss are important determinants of your credibility. If you feel that you do not currently have sufficient credibility, perhaps you can recruit someone who does to support your efforts. Most often this will be someone inside the firm, although an outsider can also be used.

Prior to making the formal presentation or submitting the authorization request, take every opportunity to share your thinking and enthusiasm with the managers who must approve the project. In addition to demonstrating your commitment to these managers, you will get an idea of their concerns and reservations, which you can then be prepared to address at the appropriate time. This type of softening also gives people time to start thinking about the project in detail, perhaps cutting down on the time required for consideration after the formal presentation.

Identify the informal chain of approval. Although some of these people may not have official sign-off authority, their opinions will count in the final decision. Softening efforts will also need to be directed at these individuals. Any objections they have will need to be taken seriously. The stronger their support, the better your chances of having the project approved.

Use methods of visualizing the technology to make it tangible. This is particularly important for software-oriented technologies. As one manager told me, "Real men don't write software"—that is, it was hard for him to convince his managers that they were really benefitting from the time and expense being devoted to software development. Animated simulations, plant trips, and videotapes were all used successfully to visualize the technology in the cases I studied.

Try to anticipate the events that might complicate or delay approval. Of course, not all such events are foreseeable, but some are, such as retirements, transfers, new product introductions, and so on. If these are anticipated, and contingency plans are in hand, the process may recover from a setback much more quickly than it otherwise would.

Finally, keep the AMT justification process separate from extraneous issues. As with the previous recommendation, this will not always be possible. Due to its importance and novelty, AMT often attracts many issues — sometimes to the point of obscuring the original intent of a project. Often this attachment is the price to be paid for support from a powerful individual. If such issues can be set aside, however, the approval process is likely to proceed more simply and rapidly.

IMPLICATIONS FOR RESEARCH

My intent has been to investigate a particular type of decisionmaking process and describe, in as much depth as possible, how it occurs in several different firms. Aside from my interest in AMT justification based on its importance for competitiveness, I also conducted this study to contribute to our knowledge of how innovation decisions are made in modern American corporations. The model that appears in Chapter 9 summarizes my observation about the five firms that I studied.

Because my research strategy for this investigation was to be deep rather than broad, I chose to investigate a relatively small number of cases in great detail. While this approach provides sufficient depth to uncover the kinds of processes I have discussed, it also has limitations. Only one type of innovation — advanced manufacturing technology — and one type of organization — relatively large industrial firms — were studied. Generalizing my findings to include other types of innovations and organizations would be premature. Even within this set of innovations and organizations, five firms are too few to be considered a representative sample.

My findings were arrived at by means of an inductive process. While it was striking to discover how pervasive these dynamics were across cases, on balance, it would be most appropriate to regard these findings as propositions or hypotheses about the nature of the innovation decision process, subject to verification in other settings.

Implications for Predicting Adoption

The study has a number of implications for prediction-oriented studies of innovation, perhaps the most important of which is the finding of the importance of proponent characteristics (e.g., solidarity, credibility, and commitment). It might be appropriate to include these variables in future models of innovation adoption, in addition to the characteristics of the firm and the innovation itself.

Due to their effects on proponent characteristics, new structural dimensions of the firm might also be investigated. For example, demographic trends leading to long-standing superior/subordinate dyads may increase levels of innovation adoption because of their effect on levels of credibility in the firm. Similarly, structural configurations may affect the likelihood of solidarity emerging around an innovation; perhaps product-based structures provide greater opportunity for this to occur than do functional arrangements. Career paths that rotate individuals across functional areas may have a similar effect. These structural dimensions would likely be most relevant for innovations such as CIM, which have strong cross-functional implications.

Finally, it may be useful to look at the functional backgrounds of senior decisionmakers. Those with more technical backgrounds will probably require less translation of the benefits of technical innovations, thus making the justification process somewhat easier for the proponents. This may serve to increase the level of technological innovation in the long run.

Comparison to Previous Work

A comparison of my findings to those of the studies reviewed in Chapter 2 will clarify the similarities and differences between AMT justification and other types of decisions, and put my work in the context of the literature.

Clearly, AMT justification involves champions (Schön 1963) and their differentiated roles (Maidique 1980). While the specific roles associated with AMT justification differ somewhat from the new product-related roles identified by Maidique, the number of these roles clearly proliferates as firms become larger and more differentiated (as Maidique's model would suggest).

The process of AMT justification as I have described it is most similar to Bower's (1970) description of the decisionmaking process associated with new facilities. This is no doubt because both of these decisions occur through the capital budgeting process. Bower's process stages of definition and impetus could be applied to AMT as well as to new facility decisions. The importance of what Bower called the "situational context" was also apparent in the decisions I studied; I found this just as difficult to generalize as Bower did.

One of the main themes in Carter's (1971) article was the loyalty to departmental priorities on the part of middle managers. Carter argued that these managers screened potential investments first on the basis of these departmental priorities, and only secondarily on the basis of the welfare of the organization as a whole. In the cases I studied, it was difficult to distinguish between loyalty to subunit versus organizational goals. This may be because the AMT proponents I studied were so thoroughly convinced that the organization needed the innovation they were proposing. As March and Simon (1958) pointed out, managers often equate the goals of their subunit with the goals of their organization; the information they receive within their subunit constantly reaffirms this equivalence. On balance, it would be fair to say that the managers I studied were convinced that their efforts would benefit their organizations as a whole, even though such conclusions may have been based on limited information.

Carter also notes that managers with limited credibility will need to demonstrate greater benefits from an investment in order to get it approved than would a manager with greater credibility. In the organizations I studied, the hurdle rates were relatively fixed — they were not adjusted based on the proponent's credibility. In the spirit of Carter's finding, however, less credible proponents would need to provide stronger and more detailed evidence of the benefits they claimed; the process, as a result, often would take much longer.

Burgelman (1983) identified two different types of linking processes: technical linking and need linking. The former is primarily driven by the existence of problems and the latter, by the existence of new technical solutions. Both types were evident in the cases I studied. While the initiatives at Temple, American Plumbing Fixtures, and MBS were largely stimulated by obvious problems, International Metals and Deftech were driven by the availability of new technology. Of course, sometimes action can only take place in the presence of both a problem and a potential technical solution. At American Plumbing Fixtures, for example, people were aware of the existence of Process X for some time before the problems at the Hancock plant made it imperative to give the process serious consideration.

Burgelman also described the necessity for champions to somehow demonstrate that "the impossible," that is, a new product technology, was indeed possible. The AMT proponents I studied often were engaged in this process. In fact, visualization activities often were oriented toward convincing others that the innovation in question could actually operate as promised. This may be a distinguishing feature of technological innovation, as opposed to the types of managerial innovations studied by Kanter (1983) or the investment decisions studied by Bower and Carter.

The potential for internal corporate ventures eventually to create a change in corporate strategy was also noted by Burgelman. While AMT clearly has the potential to effect a strategic reorientation, this did not happen in the firms I studied. Temple Laboratories ultimately did not implement the innovation it considered, and International Metals implemented only a small portion of it. The implementation of CAD at Defense Technology and MRP at Monumental Building Supply helped to support the existing strategies of those firms. Although the deployment of Process X at American Plumbing Fixtures has the greatest potential for strategic reorientation, as of this writing, APF has not yet been able to take full advantage of it.

One point agreed upon by Schön, Maidique, Carter, Bower, and Burgelman is that innovation and investment decisions generally flow from the bottom up. While each described the process somewhat differently, all of them noted that ideas most often spring from lower level organizational participants and progress through a chain of approvals before a final decision is made at the chief executive level.

While most of the cases I studied fit this pattern, International Metals was an exception. IMI's CIM initiative started at the corporate staff level and, prior to approval, did not materially involve anyone far below the level of business unit top management. The CIM advocates' initial failure was in part because the senior corporate managers were unhappy that the presentation was being made entirely by corporate staff. In addition, the frustrations experienced by the CIM advocates throughout the project were largely due to the fact that solidarity within the business units had not been achieved. It is not just coincidence that the innovation and investment decisions described by most authors operate through a bottom-up process; perhaps in order for successful innovation to occur, support needs to start at lower levels of the organization and work its way up.

It may be instructive to compare my cases with those of Bower and Burgelman on another dimension—the degree of reliance on formal analytical techniques in making decisions. Bower reports that new facility decisions are marked by extensive detailed analysis that plays a major role in whether a new facility is constructed. Burgelman, on the other hand, has found that formal analysis plays a relatively minor role in the decision to support new corporate ventures. The cases of AMT justification I studied would appear to fall between the two: while there certainly was more reliance on "the numbers" than Burgelman reports, there was less reliance than reported by Bower.

This discrepancy probably lies in the degree of experience that organizations have in each of these areas, as well as in the uncertainty inherent in each type of decision. New facility decisions are common for a manufacturing firm, and while demand forecasts are notoriously unreliable, the costs of new facilities can be fairly well anticipated. While the specifics of new facility decisions may differ, their features remain essentially the same from one decision to another. Thus when a firm is faced with such a decision, it can take advantage of its experience in making similar decisions in the past.

New venture decisions, on the other hand, provide much less opportunity to benefit from past experience. These decisions by definition involve unfamiliar industries, where the rules of the game may be so different that to rely on experience may be dangerous. And while demand forecasts for existing products that require new facilities may be unreliable, forecasts for products that do not yet exist are even more so. Finally, although the costs of new facilities are easy to determine, the costs of developing new ventures are much more difficult to estimate.

As I have noted, the AMT decisions I studied fell somewhere between these two extremes. However, there was substantial variance across cases in the degree of reliance on formal analysis and standard decision routines. While AMT was a new area for these firms, AMT decisions were often perceived as being similar to more familiar decisions, such as new machinery purchases. This was particularly true in the consideration of the robotic systems at Temple and American

Plumbing Fixtures. Once the decision was framed in a somewhat traditional manner, it could proceed within a familiar framework.

The MRP decision at MBS was seen as a combination of two familiar types of decisions: new machinery and information systems. The costs of both the hardware and software were relatively easy to calculate, while the estimation of benefits was much more difficult. Defense Technology's case was similar to MBS's in that the costs were well known but the benefits were not. The CIM effort at International Metals was the most uncertain of all the projects since neither the costs nor the benefits of CIM were known with any degree of certainty. This resulted in a decisionmaking process that relied very heavily on interpersonal and political dynamics. These observations are summarized in Table 10–1.

The final author whose findings I compare to my own is Kanter (1983). As I indicated in Chapter 2, Kanter was guided by the distinction between integration and segmentalism as contrasting modes of thought and organization. It is clear that segmentalist tendencies hampered the innovative efforts in the companies I studied. In fact, in all three of the cases where more than one subunit was involved, difficulties in reaching consensus with other subunits caused complications and delays in AMT justification. Some delay was encountered at Defense Technology when the design and manufacturing subunits tried

Table 10–1. Uncertainty and Degree of Reliance on Analytical Techniques.

Case	Technology	Certainty of Costs	Certainty of Benefits	Decision Frame	Reliance on Analysis
APF	Robotics	High	High	Machinery	High
Temple	Robotics	High	High	Machinery	High
MBS	MRP II	High	Low	Machinery/ Information system	Medium
Deftech	CAD	High	Low	Machinery/ New	Low
IMI	CIM	Low	Low	New	Low

to agree on a CAD system. MBS's people encountered even greater difficulties and delays when they tried to pursue MRP II investigation jointly with their sister company. Probably the clearest display of segmentalist thinking, however, was at International Metals, where the challenge to reach consensus across three corporate staff organizations and three business units ultimately proved too formidable for the CIM proponents.

CONCLUSION

Research on organizational decisionmaking has progressed through conceptual and empirical work with several different emphases. Some studies, such as mine, have identified a particular type of decision and described a small number of cases in great detail. Work done by such authors as Bower, Burgelman, and Kanter is also part of this tradition. Others have constructed broader theories that are intended to describe a wide range of organizational decisions. Such works include Pfeffer's (1981) description of power in organizational decisions and Cohen, March, and Olsen's (1972) garbage can theory. Studies of this type usually emphasize the similarities, rather than the differences, across different types of decisions.

A new form of investigation of organizational decisionmaking has recently emerged. This form seeks to identify the contextual variables such as organization size and environmental complexity that create the differences in decisionmaking processes that can be observed both within and between firms. A good example of this new form of theorizing is the work done by Hickson et al. (1985).

In order to truly understand the process by which decisions are made in organizations, it will be necessary for researchers to undertake a synthesis of these traditions, investigating the type of decision being made (e.g., new technology), the typical characteristics of this type of decision (e.g., novelty), the characteristics of the organization in which the decision is being made (e.g., large), and the process by which the decision takes place (e.g., political).

The goal of such research would be to map the types of decisions and organizations onto their associated decisionmaking processes. We would then be able to anticipate how the characteristics of decisions and organizations jointly determine decisionmaking processes, allowing us to anticipate how any given type of decision would be made. Only then will we have achieved the creation of an overall theory of organizational decisionmaking.

REFERENCES

Bower, J.L. 1970. *Managing the Resource Allocation Process.* Boston: Harvard Business School Press.

Burgelman, R.A. 1983. "A Process Model of Internal Corporate Venturing in the Diversified Major Firm." *Administrative Science Quarterly* 28, no. 2 (June): 223–44.

Carter, E.E. 1971. "The Behavioral Theory of the Firm and Top-Level Corporate Decisions." *Administrative Science Quarterly* 16, no. 4 (December): 413–29.

Cohen, M.D.; J.G. March; and J.P. Olsen. 1972. "A Garbage Can Model of Organizational Choice." *Administrative Science Quarterly* 17, no. 1 (March): 1–25.

Hickson, D.J.; R.J. Butler; D. Cray; G.R. Mallory; and D.C. Wilson. 1985. *Top Decisions: Strategic Decision-Making in Organizations.* San Francisco: Jossey-Bass.

Kanter, R.M. 1983. *The Change Masters: Innovation and Entrepreneurship in the American Corporation.* New York: Simon and Schuster.

Maidique, M.A. 1980. "Entrepreneurs, Champions, and Technological Innovation." *Sloan Management Review* 21, no. 2 (Winter): 59–76.

March, J.G., and H.A. Simon. 1958. *Organizations.* New York: Wiley.

Pfeffer, J. 1981. *Power in Organizations.* Marshfield, Mass.: Pitman.

Schön, D.A. 1963. "Champions for Radical New Inventions." *Harvard Business Review* 41, no. 2 (March–April): 77–86.

INDEX

Abernathy, W.J., 5, 6
Advanced manufacturing technology
    (AMT), 7–10
  American industry and, 10–11
  computer-aided design (CAD) in, 8
  computer-aided engineering (CAE) in, 9
  computer-aided manufacturing (CAM) in,
    7–8
  computer-integrated manufacturing (CIM)
    in, 10
  group technology (GT) in, 9–10
  initial investigation in, 126
  justification process and, 11–14
  manufacturing resources planning
    (MRP II) in, 9
Aerospace industry, 11
Allen-Bradley, 10, 13
American Plumbing Fixtures (APF) case,
    33, 83–94, 142
  business analysis in, 91–92
  coalition building in, 93
  corporate approval in, 90–92
  cost competition and, 86–87, 92
  credibility of management in, 90–91, 93,
    133
  environmental scanning in, 93
  exploration of innovation in, 88–90
  future of business and, 87–88

  investigating innovative processes in,
    83–85, 126
  preliminary analysis in, 92–93
  technology advances and changing
    conditions and, 85–86
  visualization in, 89–90, 92–93
Asch, S.E., 136
Automation
  islands approach to, 32
  see also headings beginning with
    Computer-
Automobile industry, 11

Bias
  case study participants and, 36–37
  investment decisions and research on,
    22–23, 28
Blumberg, M., 12
Bower, Joseph, 20–22, 28, 29, 36, 130, 133,
    142, 151, 152, 153
Brandt, R., 11, 12
Budgeting techniques
  justification process and, 13
  see also Capital allocation process
Burgelman, R.A., 23–25, 28, 29, 36, 131,
    133, 151, 152, 153
Business analysis, in American Plumbing
    Fixtures (APF) case, 91–92

157

ABOUT THE AUTHOR

James W. Dean, Jr., is an assistant professor of organizational behavior in the College of Business Administration at The Pennsylvania State University. He holds a B.A. in psychology from Catholic University and an M.S. and a Ph.D. in organizational behavior from Carnegie-Mellon University. Dean's research interests include strategic organizational decisionmaking, innovation, and advanced technology implementation. He has published in such journals as *Human Relations* and the *Journal of Applied Behavioral Science.*

DATE DUE.